Pricing and Hedging Insurance
Products in Hybrid Markets

Pricing and Hedging Insurance Products in Hybrid Markets

Jan Maximilian Widenmann

Dissertation
an der Fakultät für Mathematik, Informatik und Statistik
der Ludwig–Maximilians–Universität
München

vorgelegt von
Jan Maximilian Widenmann
aus München

München, den 25.06.2013

Bibliografische Information der Deutschen Nationalbibliothek
Die Deutsche Nationalbibliothek verzeichnet diese Publikation in der Deutschen
Nationalbibliografie; detaillierte bibliografische Daten sind im Internet über
http://dnb.d-nb.de abrufbar.
   1. Aufl. - Göttingen: Cuvillier, 2013
   Zugl.: München (LMU), Univ., Diss., 2013

978-3-95404-587-7

Erstgutachterin: Prof. Dr. Francesca Biagini
Zweitgutachter: Prof. Dr. Enrico Biffis
Tag der mündlichen Prüfung: 12.12.2013

© CUVILLIER VERLAG, Göttingen 2013
   Nonnenstieg 8, 37075 Göttingen
   Telefon: 0551-54724-0
   Telefax: 0551-54724-21
   www.cuvillier.de

978-3-95404-587-7

Eidesstattliche Versicherung

(Siehe Promotionsordnung vom 12.07.11, § 8, Abs. 2 Pkt. .5.)

Hiermit erkläre ich an Eidesstatt, dass die Dissertation von mir selbstständig, ohne unerlaubte Beihilfe angefertigt ist.

Widenmann, Jan Maximilian

Ort, Datum	Unterschrift Doktorand

Zusammenfassung

Hauptaufgaben der Versicherungsmathematik sind das Erstellen von risikoangepassten, fairen Versicherungsprämien und das Mindern der Gesamtrisikobelastung des Versicherungsunternehmens. Diese Dissertation liefert hierzu einige innovative Modelle, indem "No-Arbitrage" Preisbestimmungs- und quadratische Hedging-Methoden auf eine breite Klasse von Versicherungsverträgen angewandt wird. Die maßgebliche Annahme dabei ist, dass Versicherungsmärkte Teil eines gemischten, arbitragefreien Marktes sind, auf dem die Preis- und Hedging-Vorgaben ausgearbeitet werden können.

Die wichtigste Neuerung in Bezug auf die existierende Literatur ist das Anwenden F-doppelt stochastischer Markovketten auf aktuarielle Fragestellungen. Dies ermöglicht die Untersuchung von Versicherungsverträgen mit mehreren aufeinanderfolgenden Zustandsübergängen der versicherten Peron anhand stochastischer Intensitätsprozesse, was den Modellen eine solide Gestaltungsflexibilität verleiht.

Im Hiblick auf Prämienbestimmung wird diese Flexibilität zur Ausarbeitung von Preisschemata für Arbeitslosigkeitsversicherungsprodukte genutzt. Die Intensitätsprozesse werden dabei durch mikro- und makroökonomische stochastische Kovariablenprozesse generiert, um Einflüsse und Abhängigkeitsstrukturen innerhalb von Arbeitsmärkten zu untersuchen. Als Preisregel wird die "Real-World"-Preisformel des Benchmark-Ansatzes gewählt. Dies ermöglicht die Berücksichtigung des sogenannten $\mathbb{P}$-Numéraire Portfolios, eines globalen Indikators der Marktperformance, in den Modellen. Für unterschiedliche Spezifikationen der Intensitätsprozesse werden die Versicherungsprämien in geschlossener analytischer Form oder mit Hilfe von Monte Carlo Simulationen bestimmt. Cox's proportionales Hazardmodell wird dabei auf einen Datensatz integrierter deutscher Arbeitsmarktbiographien angewandt, um Schätzungen der Intensitätsprozesse für den deutschen Arbeitsmarkt zu erhalten. Ein wichtiger Beitrag zur bestehenden Literatur ist dabei die theoretische Verknüpfung von Cox's proportionalem Hazardmodell mit der Klasse F-doppelt stochastischer Markovketten.

Im Kontext der Risikominderung für Versicherungsunternehmen werden quadratische Hedging-Methoden auf Versicherungsprodukte angewandt. Die optimalen Handelsstrategien werden dabei anhand der Galtchouk-Kunita-Watanabe-Zerlegung der Schadenprozesse bestimmt. Eine erste Untersuchung in diesem Zusammenhang behandelt ein relevantes Thema der derzeitigen Lebensversicherungs- und Pensionskassenindustrie: die Minderung von Langlebigkeitsrisiken durch Handeln von Langlebigkeitsbonds. Hierbei werden Erwartungswert-Varianz-optimale Hedgingstrategien für mehrere Lebensversicherungsprodukte bestimmt. Die erarbeiteten Resultate werden weiter verdeutlicht, indem eine generelle affine Struktur für die Mortalitäts-Intensität angenommen wird. Eine auf numerischen Simulationen basierende Risikostudie zeigt die Eignung einer sogenannten "Gratifikations Annuität", ein neu vorgeschlagenes Lebensversicherungsprodukt, für den Lebensversicherungsmarkt. In einer zweiten Modellspezifikation werden allgemeinere, zustandsabhängige Schadenprozesse untersucht. Das beinhaltete Risiko wird optimal durch Risiko-minimierende Handelsstrategien auf die Wertpapiere des zugrunde gelegten Marktes gedeckt. Die erarbeiteten Resultate werden auch hier weiter verdeutlicht, indem eine generelle affine Struktur für die Intensitätsprozesse angenommen wird.

Zusammenfassend liefert die vorliegende Dissertation neue Erkenntnisse zu Prämienbestimmungs- und Risikominderungsansätzen für eine breite Klasse von Versicherungsprodukten. Die Intensitäts-basierten Konzepte bieten dabei einen flexiblen Rahmen, unterschiedlichste Fragestellungen der versicherungsmathematischen Praxis zu behandeln.

Summary

Main tasks in actuarial mathematics are to build risk-adjusted, fair insurance premiums and to mitigate the risk exposure of the insurance company. In this context, the present thesis provides several innovative models by applying no-arbitrage pricing and quadratic hedging approaches to a wide class of insurance contracts. The leading assumption is that insurance markets are part of a hybrid, arbitrage-free market on which pricing and hedging schemes can be elaborated.

The main novelty with respect to the existing literature is the application of $\mathbb{F}$-doubly stochastic Markov chains to actuarial problems. This allows investigating insurance contracts which cover consecutive state transitions of the insured person with stochastic intensity processes, enhancing the models with a solid design flexibility.

In the context of premium determination, this flexibility is exploited by elaborating risk-adjusted pricing schemes for unemployment insurance contracts. Here, the intensity processes are generated by micro- and macro-economic stochastic covariate processes in order to investigate influences and dependence structures within labor markets. The benchmark approaches real-world pricing formula is chosen as pricing rule. This allows incorporating the $\mathbb{P}$-numéraire portfolio, a global indicator for the performance of hybrid markets, to the models. The insurance premiums are then obtained as closed form solutions as well as by Monte Carlo simulations for different specifications of the intensities. The framework of Cox's proportional hazards model is applied to a dataset of German integrated labor market biographies in order to obtain estimates for intensity processes on the German labor market. An important contribution to the existing literature in this context is the connection of Cox's proportional hazards model with the class of $\mathbb{F}$-doubly stochastic Markov chains.

Regarding risk-mitigation for insurance companies, the focus of this thesis lies on the application of quadratic hedging schemes to insurance products. The optimal trading strategies are determined via the Galtchouk-Kunita-Watanabe decompositions of the insurance claims. A first investigation in this context addresses a relevant topic of the present life insurance and pension funds industry: the mitigation of longevity risk by trading in a longevity bond. Here, mean-variance hedging strategies are derived for several life insurance products. The obtained results are further illustrated by assuming a general affine structure for the mortality intensity process. A risk-analysis, based on numerical simulations, shows the suitability of a gratification annuity, a newly proposed life insurance product, for the life insurance market. In a second model specification, more general state-dependent insurance payment processes are investigated. The comprised risk is addressed optimally by risk-minimizing hedging strategies in the primary assets on the hybrid market. The results are again further illustrated by assuming the intensities to follow affine processes in a very general setting.

To summarize, the thesis provides a new insight to premium determination and risk-mitigation approaches for a large class of insurance products. The intensity-based concepts provide flexible frameworks for addressing different problems of actuarial practice.

Preface and Acknowledgements

This thesis has been written in partial fulfillment of the requirements for the Degree of Doctor of Natural Sciences at the Department of Mathematics, Informatics and Statistics at the University of Munich (Ludwig-Maximilians-Universität München).

The content is based on four research articles which I have co-authored since I started my duties as research assistant and graduate student at the Ludwig-Maximilians-Universität München in October 2009, see Biagini and Widenmann [14], Biagini and Widenmann [15], Biagini, Groll, and Widenmann [19], and Biagini, Rheinländer, and Widenmann [20]. My position was gratefully supported by Swiss Life Insurance Solutions AG whose main business activity also gave fruitful ideas to the projects.

My first and deepest gratitude belongs to my supervisor Francesca Biagini who guided me a long way since my university studies. She arranged the cooperation between Swiss Life Insurance Solutions AG and Ludwig-Maximilians-Universität München to offer me this position. And despite her numerous duties she always arranged time for discussing many problems of work-related but also private nature. Her mathematical expertise, combined with her warm and friendly, yet strong and determined character, made me enjoy working with her all the time.

I am also very grateful to my second assessor Enrico Biffis who not only accepted to assess my thesis on short notice but also invited me to London to discuss about the relevant parts of my work. I appreciate his informative and interesting comments and ideas.

Then I would like to thank all responsible people of Swiss Life Insurance Solutions AG who started and continued financing my research position despite of any difficulties.

Special thanks go to Thorsten Rheinländer, who invited me to London and Vienna to work on, improve and finally present my research topics and to Andreas Groll who patiently implemented the models' statistical requirements with me and became a good friend.

Thanks to the whole "Workgroup Financial Mathematics", particularly to Thilo Meyer-Brandis and Gregor Svindland who always had an open door and gave me a lot of valuable input. The whole group with its nice people helped to shorten the harder research days tremendously.

Not least I am very grateful to all my friends who stand by my side in the good and the bad times.

Finally and most importantly I want to thank my family and my girlfriend for understanding, supporting, loving and believing in me. Without your endless patience and encouragement I would have never reached this point. Thank you for everything.

Munich, June 2013   *Jan Widenmann*

Contents

List of Figures

List of Tables

List of Symbols and Abbreviations

$(\Omega, \mathcal{G}, \mathbb{P})$	Complete probability space
T	Finite time horizon $T \in (0, \infty)$
$\mathbb{F}, \mathbb{G}$	Filtrations satisfying the usual conditions with $\mathcal{F}_0 = \mathcal{G}_0 = \{\emptyset, \Omega\}$
$(X_t)_{t \in [0,T]}$	$\mathbb{R}$-valued stochastic process
$(\mathbf{X}_t)_{t \in [0,T]}$	$\mathbb{R}^N$-valued stochastic process, $N \geq 2$
$\mathbb{F}^X$	Filtration, generated by $(X_t)_{t \in [0,T]}$, i.e. $\mathcal{F}_t^X = \sigma(X_u : u \leq t)$
$\mathbb{F} \vee \mathbb{G}$	Smallest filtration, generated by $\mathbb{F}$ and $\mathbb{G}$, i.e. $\mathcal{F}_t \vee \mathcal{G}_t = \sigma(\{A : A \in \mathcal{F}_t \cup \mathcal{G}_t\})$
$\mathbb{E}[X\|\mathcal{G}]$	Expectation of X under $\mathcal{G}$ w.r.t. $\mathbb{P}$.
$\mathbb{E}_{\mathbb{Q}}[X\|\mathcal{G}]$	Expectation of X under $\mathcal{G}$ w.r.t. some probability measure $\mathbb{Q}$
$\mathbb{V}ar(X)$	Variance of a square-integrable random variable
$\mathcal{M}_0^2(\mathbb{P})$	Square integrable martingales null at $t = 0$
$\mathcal{M}_0^{\mathrm{loc}}$	Local martingales null at $t = 0$
$[\mathbf{X}, \mathbf{Y}]$	Quadratic covariation matrix of semimartingales $\mathbf{X}$ and $\mathbf{Y}$
$\langle \mathbf{X}, \mathbf{Y} \rangle$	Conditional covariation matrix of semimartingales $\mathbf{X}$ and $\mathbf{Y}$, if existing
$L(\mathbf{S})$	Predictable processes, integrable w.r.t. the semimartingale $\mathbf{S}$
$\int_{t+}^T \boldsymbol{\psi}_s^{\mathsf{T}} d\mathbf{X}_s$	$= \int \mathbb{1}_{]t,T]}(s) \boldsymbol{\psi}_s^{\mathsf{T}} d\mathbf{X}_s$ for $\boldsymbol{\psi} \in L(\mathbf{S})$, $0 \leq t < T$
$L^2(\mathbf{M})$	$= \left\{ \boldsymbol{\psi} \ \middle\| \ \boldsymbol{\psi} \text{ predictable }, \mathbb{E}\left[\int_0^T \boldsymbol{\psi}_s^{\mathsf{T}} d[\mathbf{M}]_s \boldsymbol{\psi}_s\right]^{\frac{1}{2}} < \infty \right\}$
$\mathcal{I}^2(\mathbf{M})$	$= \left\{ (\int_0^t \boldsymbol{\phi}_s^{\mathsf{T}} d\mathbf{M}_s)_{t \in [0,T]} \ \middle\| \ \boldsymbol{\phi} \in L^2(\mathbf{M}) \right\} \subseteq \mathcal{M}_0^2$
$\mathcal{P}(I)$	Power set of I
$\boldsymbol{\Psi}^{\mathsf{T}}$	Transpose of a vector or a matrix
$\mathbf{I}_N$	Identity matrix in $\mathbb{R}^{N \times N}$
a.s.	almost surely
GKW	Galtchouk-Kunita-Watanabe
IAB	Institut für Arbeitsmarkt- und Berufsforschung
i.i.d.	independent and identically distributed

ODE   Ordinary Differential Equation
PPI   Payment Protection Insurance

Introduction

Insurance in Hybrid Markets

The determination of risk-adjusted, fair insurance premiums and the management of the insurance company's risk exposure are core challenges in actuarial science. The developments on the insurance markets show the importance of well elaborated models which account for the economic environment of the insurance company. More explicitly, insurance companies

- can sell parts of their insurance risk by issuing insurance linked products on the financial markets, see Weber [95],

- can link the benefits of their insurance contracts to the performance of the assets on the stock markets by offering unit-linked insurance products, see Møller [71] or Vandaele and Vanmaele [93],

- have the possibility to invest in financial markets and hedge against their risks with financial instruments.

Insurance markets should therefore be considered as part of one big hybrid market in which appropriate pricing and risk-mitigation schemes are elaborated.

The present thesis introduces several frameworks in this context by applying no-arbitrage pricing schemes and quadratic hedging approaches to a large class of insurance contracts. All presented results are based on four research articles which have been submitted to refereed journals. The articles Biagini and Widenmann [14] as well as Biagini, Groll, and Widenmann [19] address the problem of flexibly modeling and pricing unemployment insurance contracts while the articles Biagini and Widenmann [15] as well as Biagini, Rheinländer, and Widenmann [20] cover the issues of optimally hedging insurance contracts in general settings.

A major novelty to the existing literature is to consider the underlying stochastic process, describing the insured person's progress in time of sojourning in the states, considered by an insurance policy, as an $\mathbb{F}$-doubly stochastic Markov chain. This class of stochastic processes was introduced in Jakubowski and Nieweglowski [61] and extends the classic notion of Markov chains by including a reference filtration $\mathbb{F}$, characterizing e.g. additional market information. An important property of $\mathbb{F}$-doubly stochastic Markov chains is that they may admit matrix-valued stochastic intensity processes. This allows elaborating more flexible models compared to the results of e.g. Møller [72] where a (classical) Markov chain with a deterministic matrix-valued intensity function is considered. Well known examples of $\mathbb{F}$-doubly stochastic Markov chains are the reduced-form or hazard-rate models of credit risk or life insurance, provided they admit the so called immersion property. Here, the state space consists of only two states with the second state being absorbing such that there can only occur one transition in time. There exists a vast literature on pricing and hedging within this type of models, see e.g. Barbarin [6], Biagini and Cretarola [10, 11, 12], Biagini and Schreiber [13], Biagini et al. [18, 17], Bielecki and Rutkowski [24] or Bielecki et al. [21, 22], such that the consideration of (general) $\mathbb{F}$-doubly stochastic Markov chains covers and extends most of these works to a multi-state framework. The advantage that subsequent transitions can be considered facilitates the investigation of a larger class of insurance contracts, e.g. general payment protection insurance (PPI) contracts with the insured states "disabled", "unemployed", and "deceased".

Pricing Unemployment Insurance Contracts

One of the investigated issues in this thesis is the modeling and calculation of fair insurance premiums for unemployment insurance contracts. More precisely, we consider a particular unemployment insurance product which pays a priori fixed, deterministic amounts to the insured person as soon as he gets unemployed and fulfills several other claim criteria. For example, one could think of PPI products against unemployment which are always linked to some payment obligation of the insured person. If an insured event occurs, the insurance company pays the (deterministic) installments during the respective period. Given the random claim payments of this insurance contract, we apply no-arbitrage pricing, in particular the benchmark approach with its real-world pricing formula, to determine risk-adjusted insurance premiums. The use of this approach for insurance applications is motivated as follows.

Pricing Insurance Contracts with the Benchmark Approach

Pricing of random claims has ever been one of the core subjects in both actuarial and financial mathematics and there exist various approaches for calculating (fair) prices. The actuarial way of pricing usually considers the classical premium calculation principles that consist of net premium and safety loading: if C describes a random claim which the insurance company has to pay (eventually) at time $T > 0$, then the premium $\pi(C)$ to be charged for the claim is defined by

$$\pi(C) = \underbrace{\mathbb{E}\left[\frac{C}{N_T}\right]}_{\text{net premium}} + \underbrace{A\left(\frac{C}{N_T}\right)}_{\text{safety loading}}, \tag{0.1}$$

where N is a discounting process, chosen according to actuarial judgement, see also Kull [67]. Note that the net premium is the expected value of $\frac{C}{N_T}$ with respect to the real-world (or objective) probability measure $\mathbb{P}$. Possible safety loadings are $A(\frac{C}{N_T}) = 0$ (net premium principle), $A(\frac{C}{N_T}) = a \cdot \mathbb{E}[\frac{C}{N_T}]$ (expected value principle, where $a \geq 0$), $A(\frac{C}{N_T}) = a \cdot \mathbb{V}ar(\frac{C}{N_T})$ (variance principle, where $a > 0$) or $A(\frac{C}{N_T}) = a \cdot \sqrt{\mathbb{V}ar(\frac{C}{N_T})}$ (standard deviation principle, where $a > 0$), see e.g. Rolski et al. [81]. The existence of a safety loading is justified by ruin arguments and the risk-averseness of the insurance company: the net premium principle with zero safety loading is unfavourable for the insurance company as the ruin probability of an increasing collective tends towards 50% (central limit theorem).

Widely used pricing approaches in finance base on no-arbitrage assumptions, see e.g. Black and Scholes [28] and Merton [70]. A financial market consisting of several primary assets is assumed to be in an economic equilibrium in which riskless gains out of nothing (arbitrage) by trading in the assets are impossible. A fundamental result in this context is then the essential equivalence of absence of arbitrage and the existence of an equivalent (local) martingale measure, i.e. a probability measure which is equivalent to the real-world measure $\mathbb{P}$ and according to which all assets, discounted with some numéraire process, are (local) martingales. There are different versions of this result which is often called the fundamental theorem of asset pricing (FTAP), see. e.g. Delbaen and Schachermayer [44], Delbaen and Schachermayer [45], Föllmer and Schied [53], Harrison and Pliska [57] or Kabanov and Kramkov [63].

Based on the FTAP, it can then be shown that at any time $t \in [0, T]$ an arbitrage-free price $\pi_t(C)$ of a (contingent) claim C (paid at time $T > 0$) is given by

$$\pi_t(C) := S_t^* \mathbb{E}_{\mathbb{Q}} \left[\frac{C}{S_T^*} \, \middle| \, \mathcal{G}_t \right] , \tag{0.2}$$

where $\mathbb{Q}$ is an equivalent (local) martingale measure, S^* the discounting process and $\mathbb{G} = (\mathcal{G}_t)_{t \in \mathbb{R}_+}$ the filtration which expresses the information on the market. Hence, the (new) discounted price process is assumed to follow a $(\mathbb{Q}, \mathbb{G})$-martingale.

Approaches which base on no-arbitrage assumptions are strong tools for the purpose of modeling price structures because they provide access to the powerful theory of (local) martingales. Other advantages are the dynamic description of price processes and the close connection to hedging.

From an economic point of view both the safety loading in Equation (0.1) and the change to an equivalent (local) martingale measure in Equation (0.2) express the risk-averseness of the insurance company. Moreover, there exist several works which connect actuarial premium principles with the financial no-arbitrage theory. The papers Delbaen and Haezendonck [43] and Sondermann [89] both describe a competitive and liquid reinsurance market in which insurance companies can "trade" their risks among each other. Since riskless profits shall be excluded also in this setting, the no-arbitrage theory applies and insurance premiums can be calculated by Equation (0.2). Both papers actually show that under some assumptions[1] there exist risk neutral[2] equivalent (local) martingale measures which explain premiums of the form (0.1), so that these principles provide arbitrage-free prices, too. Further papers, connecting actuarial and financial valuation principles are e.g. Kull [67] and Schweizer [86]. Note that the possibility of trading insurance contracts rather applies to the secondary market in which insurance companies trade risks through reinsurance contracts or by securitization. In particular, Equation (0.2) provides reasonable prices for the secondary market. In a competitive primary market these prices generally constitute a good benchmark as well.

---

[1] The equivalent martingale measure is required here to be structure preserving, i.e. the claim process remains a compound Poisson process under $\mathbb{Q}$.

[2] For risk-neutral martingale measures, the numéraire process S^* is chosen to be the bank account S^0 in the domestic currency.

Martingale approaches are therefore very suitable for actuarial applications and get even more important for the evaluations in the aforementioned hybrid markets. Note that due to their unsystematic part of the risk, most insurance contracts are not replicable by other instruments on the hybrid market, which implies that the market is incomplete. As a consequence, there usually exist several equivalent (local) martingale measures, corresponding to the same numéraire, that guarantee the absence of arbitrage in the market. By Equation (0.2) it is then clear that defining a premium calculation principle in the market is equivalent to choosing a numéraire and an equivalent (local) martingale measure. The usual procedure in this context is to fix some numéraire and then to search for an appropriate measure. Examples, among others, are the minimal martingale measure and the minimal entropy measure. However, several measure choices seem not to be economically reasonable for hybrid markets. Moreover, it can be shown that for several insurance linked products with random jumps, the density of the minimal martingale measure may become negative and is therefore useless in the context of pricing.

To avoid these problems, we choose the benchmark approach for our pricing issue. This approach fixes the real-world probability measure $\mathbb{P}$ and tries to determine the numéraire process, more precisely a self-financing portfolio on the assets, called the $\mathbb{P}$-numéraire portfolio, such that the discounted (or benchmarked) primary assets become local martingales or, more generally, supermartingales. The existence and uniqueness of the $\mathbb{P}$-numéraire portfolio have been shown in sufficiently general settings, see Becherer [8] or Karatzas and Kardaras [65]. The existence of the $\mathbb{P}$-numéraire portfolio then guarantees the absence of arbitrage, which is defined in a stronger way than usual. There could still exist some weak form of arbitrage in the market, which would require negative portfolios of total wealth, however. In a realistic market model, such portfolios should be impossible due to the law of limited liability. A thorough description of the benchmark approach with its real-world pricing formula and its advantages for pricing insurance contracts are given in Sections 1.1.1 and 1.2.

Application to Unemployment Insurance

Choosing the benchmark approach for pricing unemployment insurance contracts intrinsically provides a first risk-factor for the insurance premium: the $\mathbb{P}$-numéraire portfolio. In the first pricing approach which bases on the results in Biagini and Widenmann [14], we assume the underlying $\mathbb{F}$-doubly stochastic Markov chain characterizing the employment-unemployment progress of an insured person in time to be time-homogeneous. A corresponding intensity is then still random but not varying over time. More precisely, we consider an $\mathbb{F}$-doubly stochastic Markov chain which is generated by a random matrix with entries, derived from the value of the $\mathbb{P}$-numéraire portfolio at maturity. In this setting several conditional independence and distribution properties can be used to transform the insurance premium into a conditional expectation with respect to the reference filtration $\mathbb{F}$ of some closed analytic expression. The insurance premium can then be further evaluated by specifying the reference filtration more precisely. In particular, we illustrate the evaluations when the reference filtration is generated by the $\mathbb{P}$-numéraire portfolio, considered to follow a Lévy process. Moreover, we show estimation and simulation results for the case when $\mathbb{F}$ is trivial, i.e. when the $\mathbb{F}$-doubly stochastic Markov chain is a (classical) time-homogeneous Markov chain.

This first model provides interesting and reasonable results and incorporates the $\mathbb{P}$-numéraire portfolio as a risk-process in an elegant way. In a second pricing approach which is based on the results in Biagini, Groll, and Widenmann [19] we generalize this framework in order to account for the aforementioned dependencies of the model in hybrid markets. We drop the assumption on time-homogeneity but assume the underlying $\mathbb{F}$-doubly stochastic Markov chain to be generated by intensity processes which are driven by individual-related as well as micro- and macro-economic covariate processes. In this framework it is generally not possible to obtain an analytic expression for the insurance premium similar to the first framework. Instead, the insurance premiums are derived by using Monte Carlo simulations.

In order to calibrate the price for the unemployment insurance products to real data, we estimate the intensity processes using Cox's proportional hazards model, see Andersen et al. [2] and Cox [37, 38]. The data set is provided by the "Institut für Arbeitsmarkt-und Berufsforschung" (IAB), the German institute for employment research, and contains a sample of integrated labor market biographies, including the duration of employment and unemployment periods between 1975-2008 of more than 1.5 million German individuals as well as several useful socio-demographic covariates, such as age, nationality, educational level, regional details, etc. In order to reflect additional dependencies of the intensity processes to macro-economic factors, we also incorporate further covariates such as time series for the MSCI-world returns and German unemployment rates.

An advantage of using Cox's proportional hazards model is the availability of adequate implementations, see for example the R-packages corresponding to Aalen et al. [1], de Wreede et al. [42] or Jackson [60]. Technically, the implemented estimators estimate the compensator processes of multivariate counting processes which count subsequent jumps of the same kind of some unspecified multi-state switching process. The estimators in this context are based on the martingale property of the compensated counting process. The question is, if one can define characteristics for the underlying multi-state switching process such that the corresponding compensator estimators also provide estimates for (parts of) these characteristics. A well known example in this context is an underlying multi-state switching process which follows a (classical) Markov chain with deterministic matrix-valued intensity function. Here, the intensity characterizes the compensator of the corresponding counting processes and vice versa such that the obtained estimators for the (deterministic) compensator provide estimators for the intensity function as well, see Andersen et al. [2]. Yet, to the best of our knowledge, a more general relation for stochastic compensators, particularly given by Cox's proportional hazards model, has not yet been established in the literature. Based on a martingale characterization in Jakubowski and Niewęgłowski [61], we bridge this gap and show that the class of $\mathbb{F}$-doubly stochastic Markov chains is the natural candidate to be considered as the underlying multi-state switching process. This relation can analogously be applied to general multiplicative hazards models as given in Andersen et al. [2].

In order to test the obtained estimation results, we apply conventional goodness-of-fit methods. The results generally show adequate performance of the estimated model parameters. We furthermore introduce a non-standard method for testing the applicability of the obtained parameters with respect to prediction by comparing actual and simulated jump times for selected paths of the data set. The results here show good predictive power which implies robustness of the Monte Carlo simulations to compute the premiums. A conclusive sensitivity analysis of the insurance premiums also confirms these findings.

In general, both frameworks represent flexible premium determination tools for unemployment insurance products since they incorporate risk factors. Moreover, they can be easily adapted to model and estimate stochastic intensities and dependence structures in many other different applications of financial and actuarial practice.

Quadratic Hedging of Insurance Contracts

The classic form of mitigating the risk exposure of an insurance company is to buy reinsurance such that parts of the risks are taken over by another insurance company. Another way is securitization. Here, parts of the risk are combined to a package of insurance linked securities which are then sold on the financial markets, see Weber [95]. The investors in these types of securities benefit from the low correlation between most types of insurance contracts to the classical types of securities like stocks or bonds. This way the insurance linked securities provide a good potential for diversification.

The third way of mitigating an insurance company's risk exposure is to hedge parts of the risk by appropriately trading in other assets. This particularly applies if the assets are correlated to the insurance contract's benefits or their (conditional) probability of occurrence. Practical examples in this direction are unit-linked life insurance products, where benefits depend on the performance of the assets, or the aforementioned unemployment insurance products, where the occurrence of the claim payment may depend to some extend on the performance of the stock markets. Moreover, there is an ongoing discussion about the introduction of so called longevity bonds which would establish the possibility for life insurance companies and pension funds to hedge parts of their longevity risk, see e.g. Biagini and Schreiber [13] or Blake et al. [31]. Longevity bonds typically involve a publicly accessible longevity index from which the mortality intensities for a wide range of age cohorts can be derived.

As already mentioned, due to their unsystematic risk part the insurance claims in consideration are not replicable by a self-financing trading strategy such that the hybrid market is incomplete. A reasonable method for optimally choosing an investment strategy is then important to cover at least parts of the risk. Well known and elaborated approaches in this context are based on quadratic optimality criteria. In the present thesis we apply mean-variance hedging and risk-minimization to a wide class of insurance contracts. For an overview on these quadratic hedging approaches we refer to Pham [75] or Schweizer [85].

To apply quadratic hedging for insurance contracts, we assume the discounted value processes of the primary assets on the hybrid market, i.e. the hedging instruments, to be non-negative (local) martingales. This provides that the mean-variance and risk-minimizing hedging strategies are derived uniquely from the well known Galtchouk-Kunita-Watanabe (GKW-) decomposition, see Ansel and Stricker [4] or Kunita and Watanabe [68].

Mean-Variance Hedging for Life Insurance Products

In a first scenario which is based on the results in Biagini, Rheinländer, and Widenmann [20] we apply mean-variance hedging to both well known and newly introduced life insurance products by trading in longevity bonds. In particular, we consider pure endowments, i.e. contracts which pay out one unit if the insured person is alive at maturity, and term insurances, i.e. payments of one unit in case the insured person dies before the maturity of the contract. Moreover, we consider general life annuities, paying out continuous rates as long as the insured person is alive. In this context, we specify a new type of (insurance) contract which we call a gratification annuity. This insurance contract would pay increasing annuity rates, proportional to the conditional mortality probability of the insured person's own age cohort, inferred from the aforementioned longevity index. Broadly speaking, a policyholder gets gratified if the insured person is healthier or belongs to a sicker age cohort than was originally expected. The concept of a gratification annuity may also be interesting because it allows diversifying unsystematic insurance risk while transferring important parts of the systematic insurance risk to the policyholder, see also Norberg [74] and Wadsworth et al. [94] in this context. Therefore, such type of insurance contract could be interesting for the life insurance market.

The longevity bond as hedging instrument is modeled as an annuity, paying continuous rate payments proportional to the conditional survival probability, again inferable from a longevity index. There is an ongoing discussion in the literature, recommending the introduction of longevity bonds on capital markets, see e.g. Blake et al. [32]. Their appropriateness as hedging instrument for longevity risks has originally been proposed by Blake and Burrows [29].

The combined position in one of the life insurance contracts and the longevity bond also resembles various types of mortality swaps, see Dahl et al. [41] for a related concept, where the floating leg (realized mortality) is exchanged versus a fixed leg (related to some mortality projection). For a more detailed overview of the securitization of mortality risk we refer to Barrieu and Albertini [7], as well as Blake et al. [30].

Given that the underlying life history of an insured person follows an $\mathbb{F}$-doubly stochastic Markov chain with the two states "dead" and "alive", we implicitly work in the classical setting of reduced form or hazard-rate models, see Bielecki and Rutkowski [24]. We therefore use well known formulas which are specific for this two-state setting.

Under the assumption that the reference filtration $\mathbb{F}$ is generated by a one-dimensional Brownian motion W, the mean-variance hedging strategies are first calculated for a single life status and then generalized to hedging strategies for a whole portfolio of insured persons following the work of Biffis and Millossovich [26]. We remark that the GKW-decompositions obtained for the mortality claims could also be derived from the results in e.g. Barbarin [5] or Blanchet-Scalliet and Jeanblanc [33] for pure endowments, in Barbarin [5] for term insurance and in Barbarin [5] or Biagini and Cretarola [12] for general annuities. In our setting, however, we work under specific but still very general model assumptions which allow computing the GKW-decompositions explicitly. The setting furthermore allows illustrating the results for an affine specification of the mortality intensity process. This assumption is very popular in the literature about modeling mortality intensities and has been suggested for example in Biffis [25], Biffis and Millossovich [26], Dahl and Møller [40], Dahl et al. [41] or Schrager [83]. Here, we can relate the optimal hedging strategies to the solutions of well known Riccati ordinary differential equations (ODEs) and analyse the results with numerical simulations.

These simulations are carried out for two specifications of the mortality intensity, following in the first case an Ornstein-Uhlenbeck process and in the second case a Feller process. Both processes are considered to be non-mean-reverting, an assumption suggested by Luciano and Vigna [69] or Blake et al. [30]. In this context, we compare the optimal hedging strategies and their residual hedging error for a gratification annuity and a simple life annuity.

For further differences and advantages of the given framework to the ones existing in the vast literature on quadratic hedging of financial insurance derivatives, like e.g. in Barbarin [5], Dahl and Møller [40], Dahl et al. [41], Møller [71] or Møller [72], the interested reader is referred to Biagini, Rheinländer, and Widenmann [20].

Risk-Minimization for General Insurance Contracts

With similar techniques and ideas to the first hedging framework we generalize the setting and apply risk-minimization to a large class of insurance contracts, allowing also to model several consecutive state transitions of the insured person. The results here can similar be found in Biagini and Widenmann [15].

More specifically, we consider a general $\mathbb{F}$-doubly stochastic Markov chain which admits an intensity, and propose general insurance contracts as being defined by three types of insurance payments: state-dependent payments at maturity, state-dependent annuity-type payments, and (transition-dependent) payments at the transition-time from one state to another. This definition covers a large set of currently adopted insurance policies and is motivated by the definitions of rating sensitive claims in Jakubowski and Niewęgłowski [62] or defaultable claims in Bielecki et al. [23]. It covers the aforementioned insurance contracts of pure endowment, term insurance, general annuities and PPI as well as the concepts of insurance contracts, given in Møller [72] or Norberg [73].

Extending the results in Jakubowski and Niewęgłowski [62] who applied $\mathbb{F}$-doubly stochastic Markov chains in the context of replicating rating-sensitive financial claims we obtain the GKW-decomposition for the discounted value process of general insurance contracts with respect to a square-integrable $\mathbb{F}$-martingale.

In order to elaborate risk-minimizing hedging strategies it is then necessary to specify the underlying market. To this end, we assume that the reference filtration $\mathbb{F}$ is generated

by an N-dimensional Brownian motion $\mathbf{W}$ and that the assets on the hybrid market are
$\mathbb{F}$-adapted. In this setting the risk-minimizing hedging strategies are derived for general
insurance contracts with a deterministic payment structure with respect to the assets on
the market. Similar to the first framework on mean-variance hedging for life insurance
contracts, the results are then further specified within a general affine setting for the
intensity processes of the underlying $\mathbb{F}$-doubly stochastic Markov chain.

Guideline through the Thesis

Chapter 1

Chapter 1 introduces the basic notations, definitions and results which are used throughout
the thesis. In Section 1.1, the notations and definitions for hybrid markets are given based
on which the benchmark approach with its real-world pricing formula and the quadratic
hedging approaches are overviewed in Subsections 1.1.1 and 1.1.2. Section 1.2 highlights
the appropriateness of the benchmark approach and the quadratic hedging approaches for
actuarial applications and connects their general concepts and results with each other.

Chapter 2

Chapter 2 is devoted to the pricing of unemployment insurance products. In Section 2.1
the specific form of the unemployment insurance contracts in consideration is presented
based on which the corresponding insurance claim is specified. Using the real-world pricing
formula of the benchmark approach, first evaluations of fair insurance premiums are made.
In Section 2.2 a first framework for the insurance premiums within a time-homogeneous
setting of the underlying $\mathbb{F}$-doubly stochastic Markov chain is presented. The specific
results are then further illustrated within the Lévy process framework in Subsection 2.2.1
and within the classical Markov chain setting in Subsection 2.2.2. Section 2.3 provides the
second framework for evaluating the insurance premiums. Here, Cox's proportional hazards
model is connected with the class of $\mathbb{F}$-doubly stochastic Markov chains in Subsection 2.3.1.
Subsection 2.3.2 then briefly overviews the estimators which are then applied to the dataset,
described in Subsection 2.3.3. The estimation results are presented in Subsection 2.3.4 and
tested on their appropriateness through several goodness-of-fit methods in Subsection 2.3.5.
In Subsection 2.3.6, the estimates are used to evaluate the insurance premiums by Monte
Carlo simulations.

Chapter 3

Chapter 3 covers the mitigation of longevity risk by trading in a longevity bond. Here,
Section 3.1 establishes the specific modeling framework, used for deriving explicitly the
mean-variance hedging strategies. These are established for a single life status in Section
3.2 and for insurance portfolios in Section 3.3. The results are then further illustrated

within an affine structure for the mortality intensity in Section 3.4. Finally, in Section 3.5 a risk study comparing a gratification and a simple life annuity is shown.

Chapter 4

Chapter 4 considers risk-minimization for general insurance contracts by appropriately trading in the hybrid market's assets. Section 4.1 defines state-dependent insurance benefits, characterizing general insurance contracts, and illustrates the definitions according to the specific insurance products, introduced in Chapters 2 and 3. Section 4.2 provides the GKW-decompositions for the discounted intrinsic value process of a general insurance contract. In Section 4.3, risk-minimizing hedging strategies for general insurance contracts with a deterministic payout structure are specified. Similar to Section 3.4, the strategies are further illustrated within an affine structure for the intensities in Section 4.4.

Chapter 5

Chapter 5 concludes the thesis by summarizing the most important results.

1. Preliminaries on Hybrid Financial and Insurance Markets

1. Preliminaries on Hybrid Financial and Insurance Markets

In this chapter, we introduce the understanding of a hybrid market with its basic notations, definitions, and concepts. In particular we overview the benchmark approach with its real-world pricing formula for T-contingent claims and cumulative payment processes and summarize the main results on the quadratic hedging approaches "mean-variance hedging" and "risk-minimization". Finally, we underline the connection between the benchmark approach and the quadratic hedging approaches and highlight their appropriateness for actuarial applications.

1.1. The Hybrid Market

As stated in the introduction, we apply martingale pricing and quadratic hedging schemes to insurance contracts. All frameworks in this context are based on the existence of an underlying market, allowing to account for the economic environment of an insurance company. The following set-up for a (hybrid) market is standard in the literature and can e.g. very similarly be found in Biagini et al. [16].

We consider some finite time-horizon $T \in (0, \infty)$ and a frictionless financial market model in continuous time, which is set up on the complete probability space $(\Omega, \mathcal{G}, \mathbb{P})$ equipped with some filtration $\mathbb{G} = (\mathcal{G}_t)_{t \in [0,T]}$ satisfying the usual conditions, see e.g. Protter [80], with $\mathcal{G}_0 = \{\emptyset, \Omega\}$.

On the market, we can find $d + 1$ non-negative, adapted tradable (primary) security account processes, denoted by $S^j = (S^j_t)_{t \in [0,T]}$, $j \in \{0, 1, ..., d\}$, $d \geq 1$. The process S^0 is assumed to be a (locally) risk-less bank account in the domestic currency, i.e. there exists a short-rate process $r = (r_t)_{t \in [0,T]}$ such that $S^0_t = e^{\int_0^t r_s ds} > 0$, $t \in [0,T]$.

We write $\bar{\mathbf{S}} = \left((S^0_t, S^1_t, ..., S^d_t)^\intercal\right)_{t \in [0,T]}$ for the $d + 1$-dimensional random vector process, consisting of the $d + 1$ assets, and assume that $\bar{\mathbf{S}}$ is a càdlàg semimartingale.

The market is considered as a hybrid market covering financial and insurance markets. We therefore interpret the securities S^j to describe, among others, stocks but also insurance products.

Market participants can trade in the assets in order to reallocate their wealth. The meaning of trading in financial stocks is thereby as usual. However, we may need to explain what we mean by trading insurance products: on the one hand, insurance companies can sell and buy fractions of insurance claims on the secondary market. The classical form here is to offer or purchase reinsurance, see also Sondermann [89] or Delbaen and Haezendonck [43]. Another possibility is to offer or purchase securitized insurance risks on the capital markets, see Weber [95]. On the other hand, offering and terminating insurance contracts (either by the insurance company or by the policy holder) can be interpreted as trading possibilities for insurance contracts on the primary market as well. Since this understanding of trading on the primary insurance markets is a bit vague, the concept of arbitrage-free pricing rather applies to the secondary market prices. In a competitive primary insurance market, however, these prices constitute a good price benchmark as well.

We represent a trading strategy in the assets as an $\mathbb{R}^{d+1}$ valued, predictable process $\bar{\boldsymbol{\delta}} = \left((\delta_t^0, \delta_t^1, ..., \delta_t^d)^{\mathsf{T}}\right)_{t \in [0,T]} \in L(\bar{\mathbf{S}})$. Here, δ_t^j represents the units of asset j, held at time t by a market participant. The portfolio value S_t^δ at time $t \in [0, T]$ is then given by

$$S_t^\delta = \bar{\boldsymbol{\delta}}_t^{\mathsf{T}} \bar{\mathbf{S}}_t = \sum_{j=0}^d \delta_t^j S_t^j \, .$$

A strategy $\bar{\boldsymbol{\delta}} \in L(\bar{\mathbf{S}})$ is called *self-financing* if changes in the portfolio value are only due to changes in the assets and not due to in- or outflow of money, i.e. if

$$S_t^\delta = S_0^\delta + \int_{0+}^t \bar{\boldsymbol{\delta}}_s^{\mathsf{T}} \, d\bar{\mathbf{S}}_s \, , \quad t \in [0, T] \, , \tag{1.1}$$

or equivalently

$$dS_t^\delta = \bar{\boldsymbol{\delta}}_t^{\mathsf{T}} \, d\bar{\mathbf{S}}_t \, .$$

In order to prevent doubling strategies on the market, self-financing strategies are usually further limited to admissible strategies. These are self-financing strategies whose portfolio value is uniformly bounded from below. We particularly assume them to be bounded from below by zero and denote by $\mathcal{V}_x^+$ ($\mathcal{V}_x$) the set of all strictly positive (non-negative), finite and self financing portfolio values $S^\delta = (S_t^\delta)_{t \in [0,T]}$ with initial capital $S_0^\delta = x > 0$ (≥ 0).

Efficient markets are generally assumed to be arbitrage-free. Broadly speaking, an admissible portfolio $S^\delta \in \mathcal{V}_x$ is considered to be an arbitrage opportunity if its portfolio value generates some risk-less profit out of nothing. The following definition provides a strong form of arbitrage which motivates the general mathematical meaning of this concept.

Definition 1.1.1. *An admissible portfolio $S^\delta \in \mathcal{V}_x$ permits* arbitrage *if*

$$\mathbb{P}(S_\tau^\delta = 0) = 1$$

and

$$\mathbb{P}(S_\sigma^\delta > 0 \,|\, \mathcal{G}_\tau) > 0$$

for some stopping times $0 \leq \tau \leq \sigma \leq T$.

There exist several works on different concepts of absence of arbitrage on a market. Well known are the definitions of "No Arbitrage (NA)", "No Unbounded Profit with Bounded Risk (NUPBR)" and "No Free Lunch with Vanishing Risk (NFLVR)", see e.g. Delbaen and Schachermayer [44, 45], Harrison and Pliska [57], Karatzas and Kardaras [65] or Takaoka [90] for an overview.

The main reason for defining arbitrage differently is because the absence of arbitrage, defined in the respective way, guarantees the existence of some interesting objects within different settings for the market. For example if the discounted assets on the market constitute a (locally) bounded, càdlàg semimartingale, the "NFLVR" condition is equivalent to the existence of an equivalent (local) martingale measure, i.e. a probability measure $\mathbb{Q}$ which is equivalent to the real-world (or objective) probability measure $\mathbb{P}$ and according to which the discounted primary assets are (local) martingales, see Delbaen and Schachermayer [44][1].

Another example in this context, which is evident for the benchmark approach, introduced in the next subsection, is that the "NUPBR" condition is equivalent to the existence of the so called $\mathbb{P}$-numéraire portfolio, see Karatzas and Kardaras [65].

Definition 1.1.2. *A portfolio $S^{\delta^*} \in \mathcal{V}_1^+$ is called $\mathbb{P}$-numéraire portfolio if every non-negative portfolio $S^\delta \in \mathcal{V}_x$, discounted with S^{δ^*}, forms a $(\mathbb{G}, \mathbb{P})$-supermartingale. In particular, we have*

$$\mathbb{E}\left[\frac{S_\sigma^\delta}{S_\sigma^{\delta^*}} \;\middle|\; \mathcal{G}_\tau \right] \leq \frac{S_\tau^\delta}{S_\tau^{\delta^*}} \quad \mathbb{P}\text{-}a.s. \tag{1.2}$$

for all stopping times $0 \leq \tau \leq \sigma \leq T$.

Remark 1.1.3. *Every primary asset S^j, $j \in \{0, ..., d\}$ is itself the non-negative portfolio value process of the self-financing strategy $\bar{\delta}$, given by $\bar{\delta}_t = (0, ..., 0, 1, 0, ..., 0)^\intercal$, $t \in [0, T]$, with 1 at the j-th entry. Hence, Definition 1.1.2 implies that each primary asset, discounted with the $\mathbb{P}$-numéraire portfolio is a supermartingale.*

Both the "NFLVR" and the "NUPBR" condition therefore guarantees the existence of a combination of probability measure and numéraire process, such that the discounted market obtains a particular property. This combination then allows to address questions on pricing and hedging T-contingent claims and cumulative payment processes which are defined as follows.

Definition 1.1.4. *For every maturity $T \in (0, \infty)$, a T-contingent claim is a non-negative, $\mathcal{G}_T$-measurable random variable C.*

Definition 1.1.5. *A discounting process $S^* = (S_t^*)_{t \in [0,T]}$ is the portfolio value process of a strictly positive self-financing portfolio, i.e. $S^* \in \mathcal{V}_x^+$.*

Assumption 1.1.6. *The asset S^d can be substituted by the discounting process S^*. We then denote this market by $(\boldsymbol{S}, S^*)$, where $\boldsymbol{S} = (S^0, ..., S^{d-1})$ is the original market, reduced by S^d.*

---

[1]If the discounted primary assets are taken as general càdlàg semimartingales, then "NFLVR" is equivalent to the existence of an equivalent sigma-martingale measure, see Delbaen and Schachermayer [45].

Note that Assumption 1.1.6 is satisfied in most settings of practical interest. It particularly holds if S^d (or any other primary asset) is strictly positive and is taken as the discounting process S^*. Moreover it holds if the market, extended with S^*, is redundant in the sense that at any time $t \in [0, T]$, the value S_t^d can be derived from the portfolio S_t^*, and the value $\mathbf{S}_t$ of the reduced market at time t. In the following, we provide several results with respect to a (general) discounting process S^*, such that by Definition 1.1.5, these results cover several settings where $S^* = S^0$ or $S^* = S^{\delta^*}$. If we provide results for the explicit numéraires S^0 or S^{δ^*}, we will highlight this in the formulas.

Definition 1.1.7. *Given a discounting process S^*, we denote by*

$$(\widehat{\mathbf{S}}, 1) := \left(\frac{\mathbf{S}}{S^*}, 1 \right) \quad \widehat{S}^\delta := \frac{S^\delta}{S^*} \quad \widehat{C} := \frac{C}{S^*}$$

the discounted market, portfolio, or T-contingent claim, respectively.
If more particularly $S^* = S^{\delta^*}$, we call $(\widehat{\mathbf{S}}, 1)$, $\widehat{S}^\delta$, and $\widehat{C}$ the benchmarked market, portfolio or T-contingent claim.

Any T-contingent claim C generally represents a random claim payment which is due at time T. Yet, particularly in the context of insurance contracts, not only one claim payment may occur during the contract's lifetime. Also several subsequent claim payments, e.g. annuities, need to be considered. This is why we introduce now the notion of cumulative payment processes.

Definition 1.1.8. *A cumulative payment process $D = (D_t)_{t \in [0,T]}$ is a non-negative increasing $\mathbb{G}$-adapted càdlàg process.*

Remark 1.1.9. *1) As its name suggests, every cumulative payment process represents cumulative payments, i.e. for every $0 \le s < t \le T$, $D_t - D_s$ represents the sum of payments which have been paid after time s up to and including time t.*

2) The assumption that D is non-negative and increasing is only made for the sake of simplicity as all payment processes which we will model in the following chapters are of these type. More generally, one could define a cumulative payment process $\widetilde{D}$ to be a $\mathbb{G}$-adapted càdlàg process of finite variation, see e.g. Biagini and Widenmann [15] or Jakubowski and Niewęgłowski [62]. Then, $\widetilde{D}$ could represent both the cumulative incoming and outgoing payments (e.g. subsequent claim- and premium payments of an insurance contract) and could therefore become negative or non-increasing.

Example 1.1.10. *Consider a contract in which $n \in \mathbb{N}$ annuity-type payments d_i, $i \in \{1, ..., n\}$ can occur at fixed payment dates $0 \le T_1 < T_2 < ... < T_n = T$. Each payment d_i is modeled as a non-negative $\mathcal{G}_{T_i}$-measurable random variable. Then the corresponding cumulative payment process $D = (D_t)_{t \in [0,T]}$ is given as*

$$D_t = \sum_{i=1}^{n} \mathbb{1}_{[0,t]}(T_i) d_i . \tag{1.3}$$

Given a discounting process S^*, we can discount the payments d_i with the respective value $S^*_{T_i}$ and obtain the discounted cumulative payment process $\widehat{D} = (\widehat{D}_t)_{t\in[0,T]}$ as

$$\widehat{D}_t := \sum_{i=1}^n \mathbb{1}_{[0,t]}(T_i)\frac{d_i}{S^*_{T_i}} = \sum_{i=1}^n \mathbb{1}_{[0,t]}(T_i)\frac{1}{S^*_{T_i}}(D_{T_i} - D_{T_{i-1}}) = \int_0^t \frac{1}{S^*_s}dD_s \,. \qquad (1.4)$$

Equation (1.4) provides the motivation for the definition of a discounted cumulative payment process.

Definition 1.1.11. *Given a cumulative payment process D and a discounting process S^*, the discounted cumulative payment process $\widehat{D} = (\widehat{D}_t)_{t\in[0,T]}$ is defined as*

$$\widehat{D}_t := \int_0^t \frac{1}{S^*_s}dD_s \,. \qquad (1.5)$$

If $S^* = S^{\delta^*}$, then $\widehat{D}$ is called benchmarked *cumulative payment process.*

Note that the Lebesgue-Stieltjes integral in (1.5) exists, because $\frac{1}{S^*}$ is strictly positive and D by assumption increasing and càdlàg.

As indicated above, the well known no-arbitrage conditions "NFLVR" and "NUPBR" guarantee the existence of a combination $(\mathbb{Q}, S^*)$ of a probability measure $\mathbb{Q}$ and a discounting process S^* such that the discounted primary assets obtain some specific properties.

Furthermore, it can be shown that the conditions "NA" and "NUPBR" together are equivalent to the "NFLVR" condition, see e.g. Karatzas and Kardaras [65]. In particular, every (locally bounded) market model which assumes the existence of an equivalent (local) martingale measure implicitly ensures that the "NUPBR" condition holds and hence that the $\mathbb{P}$-numéraire portfolio S^{δ^*} exists. On the contrary, the existence of S^{δ^*} does not guarantee the existence of an equivalent (local) martingale measure.

The benchmark approach, which we introduce in the following subsection is only based on the assumption that the $\mathbb{P}$-numéraire portfolio exists. It defines a pricing rule, the so called real-world pricing formula which therefore provides economically reasonable prices even within market settings where no equivalent (local) martingale measure exists.

1.1.1. The Benchmark Approach

We introduce here the concepts of the benchmark approach which we use for modeling price structures, particularly for unemployment insurance contracts. All fundamental results of this approach can be found in Platen and Heath [78] for jump diffusion and Itô process driven markets and in Platen [76] for a general semimartingale market. Further works on the benchmark approach and the existence of the $\mathbb{P}$-numéraire portfolio are e.g. Becherer [8], Fernholz and Karatzas [52] , Fernholz [51], Karatzas and Kardaras [65], Platen [77] or Takaoka [90].

Based on the general setting for (hybrid) markets of the former Section, we start with establishing the key assumption for the benchmark approach.

Assumption 1.1.12. *The $\mathbb{P}$-numéraire portfolio $S^{\delta^*} \in \mathcal{V}_1^+$ exists on the market.*

Remark 1.1.13. *1) If a $\mathbb{P}$-numéraire portfolio exists, it is unique as can be easily seen with the help of the supermartingale property (1.2) and Jensen's inequality, see e.g. Becherer [8].*

2) If S^{δ^*} exists, it is equal to the "growth optimal portfolio", which is defined as the portfolio with the maximal growth-rate in the market and which satisfies several other optimality criteria, see Becherer [8], Hulley and Schweizer [58], Platen [76] or Platen and Heath [78]. In particular, the $\mathbb{P}$-numéraire portfolio outperforms any other non-negative portfolio on the long run, see Platen and Heath [78] and is therefore economically reasonable to be used as numéraire process and benchmark especially for long-dated insurance contracts.

3) The assumption on the existence of the $\mathbb{P}$-numéraire portfolio depends on the model specifications of the market but is rather weak as the existence was proven for most model specifications of nowadays practical interest, see Becherer [8], Karatzas and Kardaras [65] or Platen and Heath [78]. As indicated above, it particularly exists for all models with existing equivalent (local) martingale measure.

4) Regarding estimation of the $\mathbb{P}$-numéraire portfolio it is shown in Platen and Heath [78] that every sufficiently diversified market portfolio yields a good proxy for it. In particular, there exist several estimation frameworks for the $\mathbb{P}$-numéraire portfolio which base on data of a world stock index, e.g. the MSCI world index, see Ignatieva and Platen [59] or Platen and Heath [78].

With the existence of the $\mathbb{P}$-numéraire portfolio and the corresponding supermartingale property (1.2), arbitrage opportunities as defined in Definition 1.1.1 are excluded, see Platen [76]. There could still exist some weaker forms of arbitrage which would require to allow for negative portfolios of total wealth, however. Because of the (mostly legally established) principle of limited liability, these portfolios should be excluded in a realistic market model: a market participant generally holds a non-negative portfolio of total wealth otherwise he would have to declare bankruptcy. Assumption 1.1.12 therefore guarantees the absence of a strong form of arbitrage which is sufficient for any realistic market model.

Let us now consider two portfolios $S^{\delta} \in \mathcal{V}_x$ and $S^{\delta'} \in \mathcal{V}_y$ with

$$\widehat{S}_T^{\delta} = \frac{S_T^{\delta}}{S_T^{\delta^*}} = \frac{S_T^{\delta'}}{S_T^{\delta^*}} = \widehat{S}_T^{\delta'} \quad \mathbb{P}\text{-a.s.}.$$

Let the benchmarked portfolio process $\widehat{S}^{\delta}$ be a martingale and the benchmarked portfolio process $\widehat{S}^{\delta'}$ be a supermartingale. Then

$$\widehat{S}_t^{\delta} = \frac{S_t^{\delta}}{S_t^{\delta^*}} = \mathbb{E}\left[\left.\frac{S_T^{\delta}}{S_T^{\delta^*}}\right| \mathcal{G}_t\right] = \mathbb{E}\left[\left.\frac{S_T^{\delta'}}{S_T^{\delta^*}}\right| \mathcal{G}_t\right] \leq \frac{S_t^{\delta'}}{S_t^{\delta^*}} = \widehat{S}_t^{\delta'}, \quad t \in [0, T], \tag{1.6}$$

and in particular

$$x = S_0^\delta \leq S_0^{\delta'} = y .$$

Hence, a rational (risk-averse) investor would always invest in a benchmarked *martingale portfolio* (if it exists) and we can give the following definition of fair wealth processes, see Platen [76].

Definition 1.1.14. *A portfolio process* $S^\delta = (S_t^\delta)_{t\in[0,T]}$ *is called* fair *if its benchmarked value process* $\widehat{S}^\delta = (\widehat{S}_t^\delta)_{t\in[0,T]}$ *forms a* $(\mathbb{G}, \mathbb{P})$-*martingale.*

According to Definition 1.1.14, it is then natural to define the so called real-world pricing formula for a T-contingent claim C as follows:

Definition 1.1.15. *For a* T-*contingent claim* C, *with* $\widehat{C} = \frac{C}{S_T^{\delta*}} \in L^1(\mathbb{P})$, *the* real-world price $\pi_t(C)$ *of* C *at time* $t \in [0,T]$ *is given by*

$$\pi_t(C) := S_t^{\delta*} \mathbb{E}\left[\frac{C}{S_T^{\delta*}} \,\middle|\, \mathcal{G}_t\right] = S_t^{\delta*} \mathbb{E}\left[\widehat{C} \,\middle|\, \mathcal{G}_t\right] . \tag{1.7}$$

The corresponding benchmarked fair price process $\left(\widehat{\pi}_t(C)\right)_{t\in[0,T]} = \left(\frac{\pi_t(C)}{S_t^{\delta*}}\right)_{t\in[0,T]}$ forms a $(\mathbb{G}, \mathbb{P})$-martingale and therefore a fair wealth process according to Definition 1.1.14.

Example 1.1.16. *Let us recall Example 1.1.10, where a series of non-negative* $\mathcal{G}_{T_i}$-*measurable payments* d_i *are made at fixed payment dates* $0 < T_1 < T_2 < ... < T_n = T$. *If we consider each payment* d_i *as a* T_i-*contingent claim, then the fair price* $\pi_t(d_i)$ *of* d_i *at time* $t \in [0, T_i]$ *is given as*

$$\pi_t(d_i) = S_t^{\delta*} \mathbb{E}\left[\frac{d_i}{S_{T_i}^{\delta*}} \,\middle|\, \mathcal{G}_t\right] .$$

Now assume that each payment obligation has been settled up to and including time $t \in [0, T]$, *then we can determine the real-world price* π_t *of the remaining payments* d_i *which occur after time* t *by summing up the single real-world prices corresponding to the payments after time* t. *With the notation of the cumulative and benchmarked cumulative payment process* D *in (1.3) and (1.4), we then obtain*

$$\begin{aligned}
\pi_t &= \sum_{i=1}^n \mathbb{1}_{(t,T]}(T_i)\pi_t(d_i) = \sum_{i=1}^n \mathbb{1}_{(t,T]}(T_i) S_t^{\delta*} \mathbb{E}\left[\frac{d_i}{S_{T_i}^{\delta*}} \,\middle|\, \mathcal{G}_t\right] \\
&= S_t^{\delta*} \mathbb{E}\left[\sum_{i=1}^n \mathbb{1}_{(t,T]}(T_i) \frac{1}{S_{T_i}^{\delta*}}(D_{T_i} - D_{T_{i-1}}) \,\middle|\, \mathcal{G}_t\right] \\
&= S_t^{\delta*} \mathbb{E}\left[\int_{t+}^T \frac{1}{S_s^{\delta*}} dD_s \,\middle|\, \mathcal{G}_t\right] = S_t^{\delta*} \mathbb{E}\left[\widehat{D}_T - \widehat{D}_t \,\middle|\, \mathcal{G}_t\right] \tag{1.8}
\end{aligned}$$

Equation (1.8) motivates the following definition of the real-world price for a general cumulative payment process at time $t \in [0, T]$, after settlement of all payments up to and including time t.

Definition 1.1.17. *The real-world price $\pi_t(D)$ for a cumulative payment process $D = (D_t)_{t \in [0,T]}$ at time $t \in [0, T]$ after settlement of all payments up to and including time t is defined as*

$$\pi_t(D) := S_t^{\delta^*} \mathbb{E}\left[\int_{t+}^{T} \frac{1}{S_s^{\delta^*}} dD_s \,\bigg|\, \mathcal{G}_t \right] = S_t^{\delta^*} \mathbb{E}\left[\widehat{D}_T - \widehat{D}_t \,\bigg|\, \mathcal{G}_t \right] . \tag{1.9}$$

Remark 1.1.18. *1) Every T-contingent claim C can be considered as a cumulative payment process $D := (D_t)_{t \in [0,T]}$, defined by $D_t = C\mathbb{1}_{\{T\}}(t)$. It then follows by (1.9) and (1.7) that for $t \in [0, T)$ we have*

$$\pi_t(D) = S_t^{\bar{\delta}^*} \mathbb{E}\left[\widehat{D}_T - \widehat{D}_t \,\bigg|\, \mathcal{G}_t \right] = S_t^{\bar{\delta}^*} \mathbb{E}\left[\widehat{C}_T \,\bigg|\, \mathcal{G}_t \right] = \pi_t(C) .$$

The concept of fair pricing a payment processes payments which occur after time $t \in [0, T)$, hence, coincides with the concept of fair pricing a T-contingent claim C. For $t = T$, the concepts differ, because $\pi_T(D)$ describes the price at time T after settlement of all liabilities up to and including time T and therefore also after settlement of the T-contingent claim C. More explicitly, we have $\pi_T(D) = 0 \neq C = \pi_T(C)$.

2) Provided the market admits the equivalent (local) martingale measure $\mathbb{Q}$ for a discounting process S^*, then the arbitrage-free price $\widetilde{\pi}_t(C)$ of any T-contingent claim at time $t \in [0, T]$ is similarly defined to (1.7) as

$$\widetilde{\pi}_t(C) := S_t^* \mathbb{E}_{\mathbb{Q}}\left[\frac{C}{S_T^*} \,\bigg|\, \mathcal{G}_t \right] . \tag{1.10}$$

Moreover, the arbitrage-free price $\widetilde{\pi}_t(D)$ for a cumulative payment process D at time $t \in [0, T]$ after settlement of all payments up to and including time t is defined as

$$\widetilde{\pi}_t(D) := S_t^* \mathbb{E}_{\mathbb{Q}}\left[\int_{t+}^{T} \frac{1}{S_s^*} dD_s \,\bigg|\, \mathcal{G}_t \right] = S_t^* \mathbb{E}_{\mathbb{Q}}\left[\widehat{D}_T - \widehat{D}_t \,\bigg|\, \mathcal{G}_t \right] . \tag{1.11}$$

Similarly to Föllmer and Sondermann [54], we define the intrinsic value process of a cumulative payment process.

Definition 1.1.19. *The intrinsic value process $U^D = (U_t^D)_{t \in [0,T]}$ of a cumulative payment process D is given as*

$$U_t^D := S_t^{\delta^*} \mathbb{E}\left[\widehat{D}_T \,\bigg|\, \mathcal{G}_t \right] = S_t^{\delta^*} \mathbb{E}\left[\int_0^T \frac{1}{S_s^{\delta^*}} dD_s \,\bigg|\, \mathcal{G}_t \right] = \pi_t(D) + S_t^{\delta^*} \int_0^t \frac{1}{S_s^{\delta^*}} dD_s . \tag{1.12}$$

Note that the benchmarked intrinsic value process $\widehat{U}^D = (\widehat{U}_t^D)_{t\in[0,T]} = \left(\frac{U_t^D}{S_t^{\delta^*}}\right)_{t\in[0,T]}$ is a $(\mathbb{G}, \mathbb{P})$-martingale.

Remark 1.1.20. *1) Considering a payment process as a tradable asset, the value process U^D covers the market price of the asset, i.e. the price $\pi_t(D)$ of the residual payments after time $t \in [0,T]$, and the accumulated payments that have already occurred up to and including time t and which have been invested in the $\mathbb{P}$-numéraire portfolio at the respective payment dates. This concept can be similarly found e.g. in Duffie [48] or Jakubowski and Niewęgłowski [62].*

2) Provided the market admits an equivalent (local) martingale measure $\mathbb{Q}$ for a discounting process S^* then we define the intrinsic value process $\widetilde{U}^D = (\widetilde{U}_t^D)_{t\in[0,T]}$ of D similarly to (1.12) by

$$\widetilde{U}_t^D := S_t^* \mathbb{E}_{\mathbb{Q}}\left[\int_0^T \frac{1}{S_s^*} dD_s \,\middle|\, \mathcal{G}_t\right] = \widetilde{\pi}_t(D) + S_t^* \int_0^t \frac{1}{S_s^*} dD_s \,. \tag{1.13}$$

1.1.2. Quadratic Hedging Approaches

The following survey on quadratic hedging approaches for square-integrable T-contingent claims and square-integrable cumulative payment processes can similarly be found in Biagini and Schreiber [13], Biagini et al. [16] or Schweizer [85]. The results base on the proofs and ideas given in Bouleau and Lamberton [34], Föllmer and Sondermann [54] or Schweizer [85] for the consideration of T-contingent claims and in Barbarin [5, Chapter 4], Møller [72] or Schweizer [87] for risk-minimization for cumulative payment processes.

According to Definition 1.1.7, we denote by $(\widehat{\mathbf{S}}, 1)$ the $\mathbb{R}^{d+1}$-valued stochastic process of the discounted primary assets with respect to the discounting process S^*. By definition, the numéraire process S^* can be any strictly admissible portfolio value or primary asset. In particular, the following general outline of the quadratic hedging approaches covers settings in which the numéraire process is the (locally) risk-less bank account S^0 or the $\mathbb{P}$-numéraire portfolio S^{δ^*}.

Assumption 1.1.21. *The discounted financial market $(\widehat{\mathbf{S}}, 1)$ is a local $\mathbb{P}$-martingale.*

Remark 1.1.22. *1) If $S^* = S^0$ represents a (locally) risk-less bank account, then $\mathbb{P}$ is itself a risk-neutral equivalent (local) martingale measure.*

2) With Assumption 1.1.21 it follows by the non-negativity of all primary assets and the discounting factor S^* that $(\widehat{\mathbf{S}}, 1)$ is a supermartingale. In particular, Assumption 1.1.21 also covers several settings in which the $\mathbb{P}$-numéraire portfolio exists and $(\widehat{\mathbf{S}}, 1)$ describes the benchmarked market.

Based on Assumption 1.1.21, we would like to apply the quadratic hedging approaches *mean variance hedging* and *risk-minimization*. To this end, we assume each discounted

T-contingent claim $\widehat{C}$ and each discounted cumulative payment process $\widehat{D}$ to be square-integrable, i.e. $\mathbb{E}\left[\widehat{C}^2\right] < \infty$ and $\mathbb{E}\left[\widehat{D}_T^2\right] < \infty$.

Note again that in an incomplete hybrid market, not every square-integrable discounted T-contingent claim $\widehat{C}$ or discounted cumulative payment process $\widehat{D}$ is replicable by a self-financing portfolio[2]. The major task then is to choose a strategy within a set of trading strategies which is optimal according to some criteria. Both mean-variance hedging and risk-minimization provide unique trading strategies which are optimal according to risk that is measured in an expected quadratic sense. The set of the considered trading strategies in this context is given as follows.

Definition 1.1.23. *An L^2-strategy is a pair $(\boldsymbol{\phi}, \phi^*)$, such that $\boldsymbol{\phi} \in L^2(\widehat{\boldsymbol{S}})$ and ϕ^* is a real-valued $\mathbb{G}$-adapted process, such that the discounted portfolio value process*

$$\widehat{S}_t^\phi = \boldsymbol{\phi}_t^\intercal \widehat{S}_t + \phi_t^*, \quad t \in [0, T],$$

is right-continuous and square integrable.

Note that an L^2-strategy directly trades in the discounted assets on the market. The definition of a self-financing L^2-strategy is then analogue to (1.1) in this context.

For a square-integrable discounted T-contingent claim $\widehat{C}$, both quadratic hedging approaches have an intuitive motivation. On the one hand, the basic idea of mean-variance hedging is to consider L^2-strategies which are self-financing. The mean-variance optimal trading strategy is then the unique L^2-strategy which minimizes the expected squared distance between the discounted portfolio value and $\widehat{C}$ at time T. On the other hand, the idea of risk-minimization is to drop the self-financing property but request for strategies whose discounted portfolio values perfectly replicate $\widehat{C}$. Perfect replication can always be achieved by adding or withdrawing money to or from the portfolio. A risk-minimizing strategy then guarantees that this adding and withdrawing money is risk-optimal at each time instance in an expected squared sense. In particular, this idea can be extended to the case of a square-integrable discounted cumulative payment process $\widehat{D}$, see Møller [72]. As every square-integrable discounted T-contingent claim $\widehat{C}$ can be defined as a discounted cumulative payment process $\widehat{D}$, see Remark 1.1.18, we present the results on risk-minimization only for square integrable discounted cumulative payment processes, following the results in Møller [72].

Definition 1.1.24. *A self-financing L^2-strategy $(\boldsymbol{\varphi}, \varphi^*)$ is called* mean-variance hedging strategy *for a square-integrable discounted T-contingent claim $\widehat{C}$, if it satisfies*

---

[2]The definition of a $\widehat{D}$-replicating portfolio is e.g. given in Jakubowski and Niewęgłowski [62] or Møller [72].

$$R_0^{\widehat{C}} := \mathbb{E}\left[\left(\widehat{C} - \widehat{S}_T^{\varphi}\right)^2\right] = \mathbb{E}\left[\left(\widehat{C} - \widehat{S}_0^{\varphi} - \int_{0+}^{T} \boldsymbol{\varphi}_s^{\mathsf{T}} d\widehat{\boldsymbol{S}}_s\right)^2\right]$$

$$= \min_{\boldsymbol{\phi} \in L_{sf}^2} \mathbb{E}\left[\left(\widehat{C} - \widehat{S}_0^{\phi} - \int_{0+}^{T} \boldsymbol{\phi}_s^{\mathsf{T}} d\widehat{\boldsymbol{S}}_s\right)^2\right] \tag{1.14}$$

where L_{sf}^2 is the set of all self-financing L^2-strategies.

Remark 1.1.25. *It is well known that for every self-financing trading strategy* $(\boldsymbol{\phi}, \phi^*)$*, the process* ϕ^* *can be derived from the process* $\boldsymbol{\phi}$ *and the initial discounted portfolio value* $\widehat{S}_0^{\phi}$*, because we have*

$$\phi_t^* = \widehat{S}_t^{\phi} - \boldsymbol{\phi}_t^{\mathsf{T}} \widehat{\boldsymbol{S}}_t$$
$$= \widehat{S}_0^{\phi} + \int_{0+}^{t} \boldsymbol{\phi}_s^{\mathsf{T}} d\widehat{\boldsymbol{S}}_s - \boldsymbol{\phi}_t^{\mathsf{T}} \widehat{\boldsymbol{S}}_t.$$

Now we turn to risk-minimization. To this end, let $\widehat{D} = (\widehat{D}_t)_{t \in [0,T]}$ be a square-integrable discounted cumulative payment process. In order to define a risk-minimizing hedging strategy for $\widehat{D}$, we introduce the following processes.

For an L^2-strategy $(\boldsymbol{\phi}, \phi^*)$ the *discounted cumulative cost process* $\widehat{C}^{\phi} = (\widehat{C}_t^{\phi})_{t \in [0,T]}$ is defined by

$$\widehat{C}_t^{\phi} := \widehat{S}_t^{\phi} - \int_{0+}^{t} \boldsymbol{\phi}_s^{\mathsf{T}} d\widehat{S}_s + \widehat{D}_t,$$

describing the accumulated costs of the trading strategy[3] $(\boldsymbol{\phi}, \phi^*)$ plus the payments $\widehat{D}_t$ up to and including time t. Note that in the context of payment processes, $\widehat{S}_t^{\phi}$ should therefore be interpreted as the discounted portfolio value held at time t *after* the payments $\widehat{D}_t$ have been made, see Møller [72]. In particular, $\widehat{S}_t^{\phi}$ is the discounted value of the portfolio upon settlement of all liabilities and a natural condition is then to restrict to 0-*admissible strategies*, satisfying

$$\widehat{S}_T^{\phi} = 0 \quad \mathbb{P}\text{-a.s.}$$

The *risk process* of $(\boldsymbol{\phi}, \phi^*)$ is then given by the conditional expected value of the squared future costs

$$R_t^{\phi} = \mathbb{E}\left[\left(\widehat{C}_T^{\phi} - \widehat{C}_t^{\phi}\right)^2 \Big| \mathcal{G}_t\right], \quad t \in [0,T], \tag{1.15}$$

and is taken as a measure of the hedger's remaining risk.

---

[3]Recall that a (general) L^2-strategy needs not to be self-financing.

Definition 1.1.26. *An L^2-strategy $(\boldsymbol{\xi}, \xi^*)$ is called* risk-minimizing *for the square-integrable discounted cumulative payment process $\widehat{D}$, if for any L^2-strategy $(\boldsymbol{\phi}, \phi^*)$ such that $\widehat{S}_T^{\xi} = \widehat{S}_T^{\phi} = 0$ $\mathbb{P}$-a.s., we have*

$$R_t^{\xi} \leq R_t^{\phi} \quad \mathbb{P}\text{-a.s.}, \ t \in [0, T],$$

i.e., $(\boldsymbol{\xi}, \xi^*)$ minimizes pointwise the risk process, introduced in (1.15).

Remark 1.1.27. *Any risk-minimizing hedging strategy $(\boldsymbol{\xi}, \xi^*)$ is* mean self-financing *which means that the cumulative discounted cost process $\widehat{C}^{\xi}$ is a $(\mathbb{G}, \mathbb{P})$-martingale, see Schweizer [85].*

The key to find both a mean-variance hedging strategy for a square-integrable discounted T-contingent claim $\widehat{C}$ and a risk-minimizing hedging strategy for a square-integrable discounted cumulative payment process $\widehat{D}$ is the well-known GKW-decomposition, see Ansel and Stricker [4] or Kunita and Watanabe [68].

Definition 1.1.28. *Given a square integrable martingale $U \in \mathcal{M}^2$ and the local martingale $\widehat{\boldsymbol{S}}$, the unique GKW-decomposition for U with respect to $\widehat{\boldsymbol{S}}$ is given as*

$$U_t = U_0 + \int_{0+}^{t} (\boldsymbol{\vartheta}_s^U)^{\mathsf{T}} \, d\widehat{\boldsymbol{S}}_s + L_t^U, \quad t \in [0, T], \tag{1.16}$$

where $\boldsymbol{\vartheta}^U \in L^2(\widehat{\boldsymbol{S}})$ and $L^U \in \mathcal{M}_0^2$, strongly orthogonal to $\mathcal{I}^2(\widehat{\boldsymbol{S}})$.

Remark 1.1.29. *1) It is well-known, that $\mathcal{I}^2(\widehat{\boldsymbol{S}})$ is a closed stable subspace of $\mathcal{M}_0^2$, see Schweizer [85, Lemma 2.1.]. In particular, for every square integrable martingale U, the GKW-decomposition (1.16) always exists.*

2) The uniqueness of the GKW-decomposition (1.16) is understood up to indistinguishability, see Ansel and Stricker [3, 4], Kunita and Watanabe [68], Schweizer [85] or Schweizer [84].

For a more comprehensive discussion about the GKW-decomposition and its application for quadratic hedging we refer to Appendix B.

Since both the discounted T-contingent claim $\widehat{C}$ and the discounted cumulative payment process $\widehat{D}$ are square integrable, their discounted intrinsic value processes $\widehat{U}^C$ and $\widehat{U}^D$ of $\widehat{C}$ and $\widehat{D}$ can be decomposed by use of the GKW-decomposition (1.16) as

$$\widehat{U}_t^C := \mathbb{E}\left[\widehat{C} \,\Big|\, \mathcal{G}_t\right] = \mathbb{E}\left[\widehat{C}\right] + \int_{0+}^{t} (\boldsymbol{\vartheta}_s^C)^{\mathsf{T}} \, d\widehat{\boldsymbol{S}}_s + L_t^C, \quad t \in [0, T], \tag{1.17}$$

$$\widehat{U}_t^D := \mathbb{E}\left[\widehat{D}_T \,\Big|\, \mathcal{G}_t\right] = \mathbb{E}\left[\widehat{D}_T\right] + \int_{0+}^{t} (\boldsymbol{\vartheta}_s^D)^{\mathsf{T}} \, d\widehat{\boldsymbol{S}}_s + L_t^D, \quad t \in [0, T], \tag{1.18}$$

where $\boldsymbol{\vartheta}^C, \boldsymbol{\vartheta}^D \in L^2(\widehat{\boldsymbol{S}})$ and $L^C, L^D \in \mathcal{M}_0^2$, strongly orthogonal to $\mathcal{I}^2(\widehat{\boldsymbol{S}})$, respectively.

Theorem 1.1.30. *1) For every square-integrable discounted T-contingent claim $\widehat{C}$, there exists a unique mean-variance hedging strategy $(\boldsymbol{\varphi}, \varphi^*)$ given by*

$$\boldsymbol{\varphi}_t = \boldsymbol{\vartheta}_t^C, \tag{1.19}$$

$$\varphi_t^* = \mathbb{E}\left[\widehat{C}\right] + \int_{0+}^t (\boldsymbol{\vartheta}_s^C)^\mathsf{T} d\widehat{S}_s - (\boldsymbol{\vartheta}_t^C)^\mathsf{T}\widehat{S}_t, \quad t \in [0,T], \tag{1.20}$$

with corresponding residual risk $R_0^C = \mathbb{E}\left[(L^C)^2\right]$. Here, ϑ^C and L^C are given by (1.17).

2) For every square-integrable discounted cumulative payment process $\widehat{D}$, there exists a unique 0-admissible risk-minimizing hedging strategy $(\boldsymbol{\xi}, \xi^*)$, given by

$$\boldsymbol{\xi}_t := \boldsymbol{\vartheta}_t^D, \tag{1.21}$$

$$\xi_t^* := \widehat{U}_t^D - \widehat{D}_t - (\boldsymbol{\vartheta}^D)_t^\mathsf{T}\widehat{S}_t, \tag{1.22}$$

with discounted portfolio value process

$$\widehat{S}_t^\xi = \mathbb{E}\left[\widehat{D}_T - \widehat{D}_t \,\middle|\, \mathcal{G}_t\right] = \mathbb{E}\left[\widehat{D}_T\right] + \int_{0+}^t \boldsymbol{\xi}_s^\mathsf{T} d\widehat{S}_s + L_t^{\widehat{D}} - \widehat{D}_t, \tag{1.23}$$

discounted optimal cost process

$$\widehat{C}_t^\xi = \mathbb{E}\left[\widehat{D}_T\right] + L_t^{\widehat{D}} = C_0^\xi + L_t^{\widehat{D}},$$

and minimal risk process

$$R_t^\xi = \mathbb{E}\left[(L_T^{\widehat{D}} - L_t^{\widehat{D}})^2 \,\middle|\, \mathcal{G}_t\right], \quad t \in [0,T],$$

where $\boldsymbol{\vartheta}^D$ and L^D are given by (1.18).

Proof 1) It is easy to see that the derivation of a mean-variance hedging strategy $(\boldsymbol{\varphi}, \varphi^*)$ of $\widehat{C}$ according to Definition 1.1.24 is equivalent to projecting the random variable $\widehat{C} \in L^2(\mathbb{P})$ on the linear space, spanned by the constants in $\mathbb{R}_+$ and the stochastic integrals $\int_{0+}^T \boldsymbol{\vartheta}_s^\mathsf{T} d\widehat{S}$ for $\boldsymbol{\vartheta} \in L^2(\widehat{S})$. Hence, the uniqueness of the GKW-decomposition and Remark 1.1.25 yield the result.

2) See Föllmer and Sondermann [54] and Schweizer [85] for the single payoff case or Møller [72] and Schweizer [87] for the extension to the case of payment streams. □

Note that the discounted portfolio value process $\widehat{S}_t^\xi$ of the risk-minimizing strategy coincides with the discounted real-world or arbitrage-free prices $\widehat{\pi}_t(D)$ or $\widehat{\widetilde{\pi}}_t(D)$ of a cumulative payment process at time $t \in [0, T]$ after settlement of all payments up to and including time t as defined in (1.9) or (1.11).

Furthermore note that both approaches rely on the fact that the discounted asset prices are local martingales under the measure $\mathbb{P}$. In a more general setting, when the vector of discounted asset prices is taken to be a semimartingale under $\mathbb{P}$, one finds the price by following the *local* risk-minimization technique, see Barbarin [5, Chapter 4] or Schweizer [87]. For more information on quadratic hedging approaches we would like to refer the interested reader to the survey paper of Schweizer [85].

1.2. Benchmark Approach and Quadratic Hedging for Insurance

In this section we examine some advantages of using the benchmark approach and quadratic hedging approaches for actuarial aspects of premium determination and risk mitigation.

As already pointed out, insurance and financial markets need to be considered as part of one arbitrage-free hybrid market. In this context, the standard actuarial premium principles defined by the risk premium loaded with some safety margin are often considered to be rather ad hoc approaches which a priori do not take into account the insurance company's economic environment. Well established no-arbitrage methods address this issue intrinsically and may hence be rather suitable for premium determination, in particular for hybrid insurance products. Actually, most of the no-arbitrage frameworks also cover standard actuarial premium principles in the sense that there exists some risk-neutral equivalent (local) martingale measure such that the price structures with respect to this measure coincide with the prices obtained by the standard actuarial premium principles, see Delbaen and Haezendonck [43], Kull [67], Schweizer [86] or Sondermann [89].

As we choose the benchmark approach with its real-world pricing formulas (1.7) or (1.9) for determining premiums of unemployment insurance contracts, we are therefore working in a more realistic setting for hybrid markets.

Another advantage of the benchmark approach is that it provides economically reasonable prices for financial and insurance products even in market set-ups where the usual risk-neutral pricing frameworks fail to work. This is due to the aforementioned fact that the existence of the $\mathbb{P}$-numéraire portfolio, the leading assumption of the benchmark approach, may be given (because the market satisfies the "NUPBR" condition) even if no risk-neutral equivalent (local) martingale measure exists (because the "NFLVR" condition fails). More specifically, given the existence of the $\mathbb{P}$-numéraire portfolio, there exists a risk-neutral equivalent (local) martingale measure $\mathbb{P}^*$ if its density process $\Lambda = (\Lambda_t)_{t \in [0,T]}$ with $\Lambda_t = \frac{\widehat{S}_t^0}{\widehat{S}_0^0}$ exists. In this case, the real-world price of a T-contingent claim or cumulative payment process coincides with the risk-neutral price with respect to the equivalent (local) martingale measure $\mathbb{P}^*$, see Platen and Heath [78]. Because most of the actuarial pricing principles can be embedded in the risk-neutral framework as well, the benchmark approach can be considered as a similar, but more reasonably justified, choice of actuarial premium principle.

A further similarity to the classic forms of actuarial pricing is the direct use of the real-world (or objective) probability measure $\mathbb{P}$. This provides huge benefits for the statistical investigations within these models. In Section 2.3 we will provide estimators for employment and unemployment intensities which can be immediately used for the premium evaluations. If, instead, we worked out the pricing formula in a risk-neutral framework, i.e. with respect to some equivalent (local) martingale measure, in order to use the estimated intensities, they may need to be adjusted according to some transformation, see Biffis et al. [27].

A further advantage is the fact that we take directly into account the role of investment opportunities in assessing premiums and reserves. The $\mathbb{P}$-numéraire portfolio is a direct and intuitive global indicator of (hybrid) market performance and dependence structure. This is particularly relevant for insurance contracts depending on the performance of financial markets and macro-economic factors, as e.g. unit-linked or unemployment insurance contracts. On the contrary, the choice and investigation of a particular (local) martingale measure for actuarial applications appears quite artificial.

Moreover, the $\mathbb{P}$-numéraire portfolio shows several optimality criteria, e.g. it maximizes the expected logarithmic utility or it outperforms any other non-negative portfolio in the long run. Insurance companies which offer long-dated insurance contracts are exposed to risk in a distant future. In order to address this risk properly, insurance companies should invest in the $\mathbb{P}$-numéraire portfolio in order to reallocate their wealth optimally. This, however, suggests to derive the insurance premiums by means of the benchmark approach and its real-world pricing formulas (1.7) and (1.9).

There is also an intrinsic relation between the quadratic hedging approaches, as introduced in Section 1.1.2, and the use of the benchmark approach. If we assume that the discounted hybrid market $(\widehat{\mathbf{S}}, 1)$ is the benchmarked market, i.e. that the primary assets are benchmarked with the $\mathbb{P}$-numéraire portfolio S^{δ^*}, then Theorem 1.1.30 states on the one hand that the unique mean-variance hedging strategy for a benchmarked T-contingent claim $\widehat{C}$ is given by $(\boldsymbol{\varphi}, \varphi^{\delta^*})$ with $\boldsymbol{\varphi}_t = \boldsymbol{\vartheta}_t^C$ and $\varphi_t^{\delta^*} = \mathbb{E}[\widehat{C}] + \int_{0+}^t (\boldsymbol{\vartheta}_t^C)^\intercal d\widehat{S}_s - (\boldsymbol{\varphi}_t^C)^\intercal \widehat{S}_t$, $t \in [0, T]$, where $\boldsymbol{\vartheta}^C$ is given by the GKW-decomposition (1.17). We then have

$$\widehat{S}_0^\varphi = \boldsymbol{\varphi}_0^\intercal \widehat{S}_0 + \varphi_0^{\delta^*} = \mathbb{E}[\widehat{C}] \ .$$

Hence, in order to initiate the self-financing mean-variance hedging strategy for $\widehat{C}$, we need to invest the same amount of money which we obtain by selling the claim $\widehat{C}$ to the price, determined by the real-world pricing formula (1.7).

On the other hand, Theorem 1.1.30, more explicitly Equation (1.23), shows that the benchmarked portfolio value process $\widehat{S}^\xi$ of the risk-minimizing hedging strategy for any square-integrable cumulative payment process $\widehat{D}$ coincides with $(\widehat{\pi}(D)_t)_{t \in [0,T]}$, the benchmarked real-world price process as given in (1.9). Therefore, the risk-minimizing strategy needs to be adjusted at any time instance such that the portfolio value meets the real-world price of the remaining payments at that time. For similar results on the relation between actuarial valuation principles and mean-variance hedging we also refer to Schweizer [86]. Also see Fontana and Runggaldier [55] for a discussion on the relation among real-world pricing, upper-hedging pricing and utility indifference pricing.

Recall that we assumed the discounted or benchmarked primary assets $\widehat{\mathbf{S}}$ to be a local martingale. For a more thorough investigation on the connection of the benchmark approach with quadratic hedging approaches, we refer to Biagini [9] and Biagini et al. [16].

To end this section we point out some further useful properties of quadratic hedging approaches towards actuarial applications. As we have seen, both the mean-variance hedging strategy and the risk-minimizing hedging strategy are determined uniquely by the GKW-decomposition (1.16). This decomposition divides the risk related to an insurance claim in a natural way. Given the GKW-decomposition with respect to the benchmarked assets $\widehat{\mathbf{S}}$ as in (1.16), the insurance risk is split into a hedgeable and some unhedgeable part. The hedgeable part can be covered by investing in a self-financing strategy and therefore bears no risk for the insurance company whereas the unhedgeable part needs to be investigated more thoroughly. In most cases the unhedgeable part can then be further decomposed into a purely unsystematic risk part related to the insured person and some residual part of systematic risk. It is a common fact that an insurance portfolio's unsystematic part of the risk can then be reduced through diversification by increasing the number of insured persons.

More specifically, we will see in the frameworks of Chapters 3 and 4 that within the $\mathbb{F}$-doubly stochastic setting the derivation of quadratic hedging strategies can be done by first decomposing the insurance risk into a stochastic integral with respect to a square-integrable $\mathbb{F}$-martingale and a stochastic integral with respect to an individual-related (local) martingale, which is given by the $\mathbb{F}$-doubly stochastic Markov chain. The stochastic integral with respect to the individual-related (local) martingale then represents the unsystematic risk, whereas the stochastic integral with respect to the $\mathbb{F}$-martingale represents the systematic risk. Depending on the respective assumptions on the filtration $\mathbb{F}$, parts of the systematic risk can be hedged by self-financing trading strategies, whereas other parts remain in the systematic risk exposure of the insurance company, see also Biagini and Schreiber [13].

Note that in the context of risk-minimization it seems as if the insurance risk is "hedged" perfectly such that one may argue that there is no risk for the insurance company. This, however, is not the case as the readjustment of the (non self-financing) trading strategy basically represents the risk for the insurance company: due to limited credit lines it is in reality not possible to put an arbitrary amount of money into the portfolio in order to address some risk exposure perfectly.

Therefore, quadratic hedging provides on the one hand recommendations for trading strategies which shall address a risk exposure in a risk-optimal way, on the other hand the GKW-decompositions of the insurance risk provide an important insight into unsystematic and systematic as well as hedgeable and unhedgeable parts of the risk. All these issues are highly relevant for nowadays actuarial practice.

2. Pricing of Unemployment Insurance Products

2. Pricing of Unemployment Insurance Products

In this chapter we introduce two general pricing schemes for unemployment insurance contracts. Given the contractual parameters of the insurance policies, a challenging problem in practice is to derive flexible premiums which can be calibrated to several risk-factors. In this context, the class of $\mathbb{F}$-doubly stochastic Markov chains, as introduced in Appendix A, allows the elaboration of appropriate frameworks because of the variety in which intensities can be modeled.

For both pricing schemes we therefore assume the underlying process of an insured person of being employed or unemployed as a two-state $\mathbb{F}$-doubly stochastic Markov chain. In the first framework we additionally assume the $\mathbb{F}$-doubly stochastic Markov chain to be time-homogeneous. Using the benchmark approaches real-world pricing formula and a particular structure of the random intensity matrix, we then obtain the price of an unemployment insurance contract in closed analytic form depending on the development of the $\mathbb{P}$-numéraire portfolio.

In the second framework, we increase the set of risk-factors and consider the more realistic case of stochastic intensity processes generating the $\mathbb{F}$-doubly stochastic Markov chain. Given the model specifications of Cox's proportional hazards model, we estimate covariate effects on the intensities on a large dataset of German labor market biographies. Using again the benchmark approach and the $\mathbb{F}$-doubly stochastic Markov chain assumption, we obtain the insurance premiums by applying Monte Carlo simulations.

Large parts of this chapter are based on the findings in Biagini and Widenmann [14] and in Biagini, Groll, and Widenmann [19].

2.1. Unemployment Insurance Contracts

In order to price unemployment insurance contracts, we perform a single insurance evaluation. This allows adjusting the insurance premium according to both individual characteristics of the insured person and micro- and macro-economic influences. This approach requires a thorough description of the insurance contracts' structures.

The basic idea of unemployment insurance is to reduce the financial deficiencies which an unemployed insured person is exposed to. In this context, we investigate insurance contracts with deterministic, a priori fixed claim payments $c_1, ..., c_N$ at predefined payment dates $0 \leq T_1 < ... < T_N = T$, $N \in \mathbb{N}$, provided the insured person is unemployed and fulfills further contractual claim criteria. Hence, these contracts can be interpreted as an annuity during an unemployment period. Note that the randomness of the claims is only due to their occurrence and not to their amount.

As a practical example one could think of PPI products against unemployment which are linked to some payment obligation of an obligor to its creditor. The claim amount is hereby defined by the payment obligation's installments which are paid according to a predefined plan. Therefore the potential claim payments for the insurance company can be considered as constants known at the contract's beginning. One may argue that obligors sometimes adjust their installment plan and that the claims are therefore stochastic, but most of the PPI contracts prevent this strategy in their exclusion clauses and rather offer an adjustment of the insurance premium according to the new, deterministic installment plan.

Further details of the considered insurance contracts which are important for the model specifications are the following.

- Regarding the method of premium payment, one differentiates between single rates where the whole insurance premium is paid at the beginning of the contract, and periodical rates. For our modeling purpose, we focus on calculating single premiums. This is motivated by PPI unemployment products which are often sold as an add-on directly by the creditor. The insurance company then receives a single rate from the creditor who in turn allocates this rate to the installments.

- In order to conclude the insurance contract, the prospective buyer must have been employed at least for a certain period before the beginning of the contract. Therefore, we only consider employed insured persons at time $t = 0$.

We also consider three time periods that belong to the exclusion clauses of the contracts and impact the insurance premium.

- The *waiting period* starts with the beginning of the contract. If an insured person becomes unemployed at any time of this period, he is not entitled to receive any claim payments during this stage of unemployment.

- The *deferment period* starts with the first day of unemployment. An insured person is not entitled to receive claim payments until the end of this period.

- The third period is comparable to the waiting period and is called the *requalification period*. The difference between the waiting and the requalification period is their beginning. The waiting period starts with the beginning of the contract and the requalification period with the end of any unemployment period that occured during the contract's duration. If an insured person becomes unemployed (again) at any time of the requalification period, he is not entitled to receive any claim payment during this stage of unemployment.

For existing unemployment insurance contracts the waiting, deferment, and requalification periods currently vary from zero to twelve months.

In order to describe this type of unemployment insurance contract mathematically, we use the definitions and assumptions of Chapter 1. More explicitly, we consider the complete, filtered probability space $(\Omega, \mathcal{G}, \mathbb{G}, \mathbb{P})$ with $\mathbb{G} = (\mathcal{G}_t)_{t \in [0,T]}$ satisfying the usual conditions with $\mathcal{G}_0 = \{\emptyset, \Omega\}$. The maturity $T = T_N$ is fixed as the final payment date of the insurance contract. The individual progress of an insured person of being employed and/or unemployed is modeled as a right-continuous, $\mathbb{G}$-adapted stochastic process $X = (X_t)_{t \in [0,T]}$ with state-space $\{1, 2\}$. If $X_t = 1$ for some $t \in [0, T]$, the insured person is assumed to be employed at that time whereas for $X_t = 2$ he is unemployed.

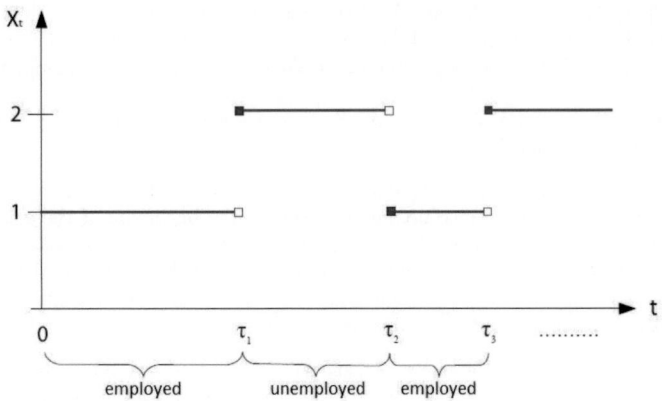

Figure 2.1.: Trajectory of a right-continuous stochastic process X with state-space $\{1, 2\}$ and jump times $\tau_0, \tau_1, \ldots$. The stochastic process is interpreted as the employment-unemployment progress of a person.

Of course, the relevant parameters for the claim payments are then the sojourn-times of the insured person in the respective states 1 and 2. To investigate these periods, we introduce the random jump times $(\tau_n)_{n\geq 0}$ of X by

- $\qquad \tau_0 := 0$

- $\qquad \tau_n := \inf\{\tau_{n-1} < t \leq T : X_t \neq X_{\tau_{k-1}}\}, \quad n \geq 1\,.$ \hfill (2.1)

Figure 2.1 shows an exemplary path of X with its respective jump times.

We denote by c_i the deterministic claim amount which is (eventually) paid at time T_i, $i = 1, \ldots, N$, and define $W, D, R \in \mathbb{R}_+$ to be the waiting, deferment, and requalification period, respectively. We think of $t = 0$ as the beginning of the contract. As we only consider insured persons which are employed at the contract's beginning we set $X_0 = 1$ $\mathbb{P}$-a.s.

How do the random claim costs of the unemployment insurance contract arise? According to the contractual specifications provided above, the insurance company has to pay the claim to the amount of c_i at time T_i if the following conditions are satisfied:

- $W < \tau_1 \leq T_i - D$
  The first jump τ_1 to unemployment of the insured person must occur after the waiting period W. Moreover, at least the deferment period D must lie between τ_1 and the payment date T_i, i.e. $T_i - \tau_1 \geq D$.
  and

- $T_i < \tau_2$

The insured person must not have jumped back to employment before the payment date T_i.

OR for $j \geq 2$

- $\tau_{2j-1} - \tau_{2j-2} > R$
  At least the requalification period R must lie in between a jump τ_{2j-2} to employment and the next jump τ_{2j-1} back to unemployment.
  and

- $W < \tau_{2j-1} \leq T_i - D$
  Any jump to unemployment τ_{2j-1} must occur after the waiting period W. Moreover, at least the deferment period D must lie between τ_{2j-1} and the payment date T_i, i.e. $T_i - \tau_{2j-1} \geq D$.
  and

- $\tau_{2j} > T_i$
  Before the payment date T_i the insured person must not have jumped back to employment.

Based on this insight, the random insurance claim C_i at the payment date T_i can be defined as

$$C_i := c_i \mathbb{1}_{\{W < \tau_1 \leq T_i - D, \tau_2 > T_i\} \cup \bigcup_{j=2}^{\infty} \{\tau_{2j-1} - \tau_{2j-2} > R, W < \tau_{2j-1} \leq T_i - D, \tau_{2j} > T_i\}} . \tag{2.2}$$

In the following, we denote

$$A_j^x := \{\tau_j \leq x\}, B_j^x := \{x < \tau_j\}, D_j^x := \{\tau_j - \tau_{j-1} > x\}.$$

Note that due to the definition of the jump times, the events $A_{2j-1}^{T_i-D} \cap B_{2j-1}^W \cap B_{2j}^{T_i}$, $j \geq 1$, are disjoint. Therefore, we can rewrite the random insurance claim C_i in (2.2) as

$$C_i = c_i \left(\mathbb{1}_{B_1^W \cap A_1^{T_i-D} \cap B_2^{T_i}} + \sum_{j=2}^{\infty} \mathbb{1}_{D_{2j-1}^R \cap B_{2j-1}^W \cap A_{2j-1}^{T_i-D} \cap A_{2j}^{T_i}} \right). \tag{2.3}$$

As X is assumed to be $\mathbb{G}$-adapted, all jump times τ_j, $j \geq 0$, are $\mathbb{G}$-stopping times. It follows then by (2.3) that C_i is a T_i-contingent claim according to Definition 1.1.4.

With the real-world pricing formula (1.7), the real-world price $\pi_t(C_i)$ of C_i at time $t \in [0, T_i]$ is then given as

$$\pi_t(C_i) = S_t^{\delta^*} c_i \left\{ \mathbb{E}\left[\widehat{\mathbb{1}}_{B_1^W \cap A_1^{T_i-D} \cap B_2^{T_i}} \,\Big|\, \mathcal{G}_t \right] + \sum_{j=2}^{\infty} \mathbb{E}\left[\widehat{\mathbb{1}}_{D_{2j-1}^R \cap B_{2j-1}^W \cap A_{2j-1}^{T_i-D} \cap B_{2j}^{T_i}} \,\Big|\, \mathcal{G}_t \right] \right\},$$

where we used the notation for the benchmarked payoff introduced in Definition (1.1.7).

Similar to Example 1.1.10 we can now sum up over all payment dates T_i to obtain the insurance premium π_t at time $t \in [0, T]$ *after* settlement of all payments up to and including time t as

$$\pi_t = \sum_{i=1}^{N} \mathbb{1}_{(t,T]}(T_i) S_t^{\delta^*} c_i \left\{ \mathbb{E}\left[\hat{\mathbb{1}}_{B_1^W A_1^{T_i-D} B_2^{T_i}} \,\Big|\, \mathcal{G}_t \right] + \sum_{j=2}^{\infty} \mathbb{E}\left[\hat{\mathbb{1}}_{D_{2j-1}^R B_{2j-1}^W A_{2j-1}^{T_i-D} B_{2j}^{T_i}} \,\Big|\, \mathcal{G}_t \right] \right\}. \quad (2.4)$$

Note that we suppress the symbol "$\cap$" for notational convenience.

Remark 2.1.1. *Similar to Example 1.1.10, we can define a cumulative payment process $D = (D_t)_{t \in [0,T]}$ for the unemployment insurance contract as*

$$D_t := \sum_{i=1}^{N} \mathbb{1}_{[0,t]}(T_i) C_i. \quad (2.5)$$

According to 1.1.11, its benchmarked representative $\hat{D} = (\hat{D}_t)_{t \in [0,T]}$ is then given as

$$\hat{D}_t = \int_{[0,t]} \frac{1}{S_u^{\delta^*}} dD_u = \sum_{i=1}^{N} \mathbb{1}_{[0,t]}(T_i) \frac{C_i}{S_{T_i}^{\delta^*}}.$$

With the real-world pricing formula (1.9) for cumulative payment processes we then obtain for $t \in [0, T]$ that

$$\pi_t(D) = \mathbb{E}\left[\int_{t+}^{T} \frac{1}{S_u^{\delta^*}} dD_u \,\Big|\, \mathcal{G}_t \right]$$

$$= \sum_{i=1}^{N} \mathbb{1}_{(t,T]}(T_i) S_t^{\delta^*} c_i \left\{ \mathbb{E}\left[\hat{\mathbb{1}}_{B_1^W A_1^{T_i-D} B_2^{T_i}} \,\Big|\, \mathcal{G}_t \right] + \sum_{j=2}^{\infty} \mathbb{E}\left[\hat{\mathbb{1}}_{D_{2j-1}^R B_{2j-1}^W A_{2j-1}^{T_i-D} B_{2j}^{T_i}} \,\Big|\, \mathcal{G}_t \right] \right\}.$$

in accordance with (2.4).

In the following, we write $\pi_t(D)$ for the (real-world) unemployment insurance premium at time $t \in [0, T]$ after settlement of all payments up to and including time t.

Because the waiting period W is a fixed period at the contracts beginning, we further distinguish the cases

1) $t < W$, and

2) $t \geq W$.

We begin by considering situation 1) and evaluate the insurance premium for this case more explicitly. We introduce first two illustrative examples which give some intuition about how to compute the terms appearing in (2.4). The proof of the subsequent proposition is then straightforward and therefore omitted.

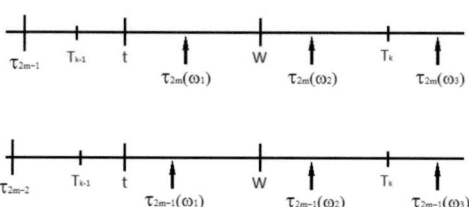

Figure 2.2.: Illustration of different scenarios of the jump times before and after time $t < W$. In the upper plot the insured person is employed at time t. In the lower plot the insured person is unemployed at time t.

Example 2.1.2. *We assume that the jump τ_{2m-2} has already occurred up to time t for some $m \geq 2$. Hence, the insured person is employed at time t. The lower plot of Figure 2.2 illustrates this scenario for three exemplary realizations $\omega_1, \omega_2, \omega_3 \in \Omega$.*

Since in this situation

- *τ_{2m-2} is known at time t and*

- *the restriction $\tau_{2j-1} > W$ is impossible to be satisfied for $j \leq m-1$,*

we obtain the insurance premium $\pi_t(D)$ at time $t \in [0, W)$ as

$$\pi_t(D) = \sum_{i=1}^{N} \mathbb{1}_{(t,T]}(T_i) S_t^{\delta^*} c_i \left\{ \mathbb{E}\left[\widehat{\mathbb{1}}_{B_{2m-1}^{W \vee (R+\kappa)} A_{2m-1}^{T_i-D} B_{2m}^{T_i}} \,\Big|\, \mathcal{G}_t \right]\Bigg|_{\kappa=\tau_{2m-2}} \right.$$
$$\left. + \sum_{j=m+1}^{\infty} \mathbb{E}\left[\widehat{\mathbb{1}}_{D_{2j-1}^{R} B_{2j-1}^{W} A_{2j-1}^{T_i-D} B_{2j}^{T_i}} \,\Big|\, \mathcal{G}_t \right] \right\}.$$

Example 2.1.3. *Assume that for some $m \geq 2$ the jumps $\tau_1, ..., \tau_{2m-1}$ have occurred up to time t. The upper plot of Figure 2.2 illustrates this scenario for three exemplary realizations $\omega_1, \omega_2, \omega_3 \in \Omega$.*

In this case, the restriction $\tau_{2j-1} > W$ is impossible to be satisfied for $j \leq m$ and the calculation of $\pi_t(D)$ boils down to

$$\pi_t(D) = \sum_{i=1}^{N} \mathbb{1}_{(t,T]}(T_i) S_t^{\delta^*} c_i \sum_{j=m+1}^{\infty} \mathbb{E}\left[\widehat{\mathbb{1}}_{D_{2j-1}^{R} B_{2j-1}^{W} A_{2j-1}^{T_i-D} B_{2j}^{T_i}} \,\Big|\, \mathcal{G}_t \right].$$

We obtain the following proposition.

Proposition 2.1.4. *If $t < W$, the insurance premium $\pi_t(D)$ at time $t \in [0, W)$ is given by*

$$\pi_t(D) = \sum_{i=1}^{N} \mathbb{1}_{(t,T]}(T_i) S_t^{\delta^*} c_i \Bigg\{ \mathbb{1}_{\{t<\tau_1\}} \left(\mathbb{E}\left[\widehat{\mathbb{1}}_{B_1^W A_1^{T_i-D} B_2^{T_i}} \,\Big|\, \mathcal{G}_t \right] + \sum_{j=2}^{\infty} \mathbb{E}\left[\widehat{\mathbb{1}}_{D_{2j-1}^R B_{2j-1}^W A_{2j-1}^{T_i-D} B_{2j}^{T_i}} \,\Big|\, \mathcal{G}_t \right] \right)$$

$$+ \mathbb{1}_{\{\tau_1 \leq t < \tau_2\}} \sum_{j=2}^{\infty} \mathbb{E}\left[\widehat{\mathbb{1}}_{D_{2j-1}^R B_{2j-1}^W A_{2j-1}^{T_i-D} B_{2j}^{T_i}} \,\Big|\, \mathcal{G}_t \right]$$

$$+ \sum_{m=2}^{\infty} \mathbb{1}_{\{\tau_{2m-2} \leq t < \tau_{2m-1}\}} \left(\mathbb{E}\left[\widehat{\mathbb{1}}_{B_{2m-1}^{W\vee(R+\kappa)} A_{2m-1}^{T_i-D} B_{2m}^{T_i}} \,\Big|\, \mathcal{G}_t \right] \Bigg|_{\kappa=\tau_{2m-2}} \right.$$

$$\left. + \sum_{j=m+1}^{\infty} \mathbb{E}\left[\widehat{\mathbb{1}}_{D_{2j-1}^R B_{2j-1}^W A_{2j-1}^{T_i-D} B_{2j}^{T_i}} \,\Big|\, \mathcal{G}_t \right] \right)$$

$$+ \sum_{m=2}^{\infty} \mathbb{1}_{\{\tau_{2m-1} \leq t < \tau_{2m}\}} \sum_{j=m+1}^{\infty} \mathbb{E}\left[\widehat{\mathbb{1}}_{D_{2j-1}^R B_{2j-1}^W A_{2j-1}^{T_i-D} B_{2j}^{T_i}} \,\Big|\, \mathcal{G}_t \right] \Bigg\} . \tag{2.6}$$

We now consider situation 2), where $t \geq W$. In this case, we have

$$\{\tau_j > t\} \subseteq \{\tau_j > W\} ,$$

such that the latter restriction is automatically fulfilled for every $\omega \in \{\tau_j > t\}$. Taking account for this and with similar arguments as in Examples 2.1.2 and 2.1.3, we obtain the following proposition.

Proposition 2.1.5. *If $t \geq W$, the insurance premium $\pi_t(D)$ at time $t \in [0,T]$ after settlement of all payments up to and including time t is given by*

$$\pi_t(D) = \sum_{i=1}^{N} \mathbb{1}_{(t,T]}(T_i) S_t^{\delta^*} c_i \Bigg\{ \mathbb{1}_{\{t<\tau_1\}} \left(\mathbb{E}\left[\widehat{\mathbb{1}}_{A_1^{T_i-D} B_2^{T_i}} \,\Big|\, \mathcal{G}_t \right] + \sum_{j=2}^{\infty} \mathbb{E}\left[\widehat{\mathbb{1}}_{D_{2j-1}^R A_{2j-1}^{T_i-D} B_{2j}^{T_i}} \,\Big|\, \mathcal{G}_t \right] \right)$$

$$+ \mathbb{1}_{\{\tau_1 \leq t < \tau_2\}} \left(\mathbb{1}_{B_1^W A_1^{T_i-D}} \mathbb{E}\left[\widehat{\mathbb{1}}_{B_2^{T_i}} \,\Big|\, \mathcal{G}_t \right] + \sum_{j=2}^{\infty} \mathbb{E}\left[\widehat{\mathbb{1}}_{D_{2j-1}^R A_{2j-1}^{T_i-D} B_{2j}^{T_i}} \,\Big|\, \mathcal{G}_t \right] \right)$$

$$+ \sum_{m=2}^{\infty} \mathbb{1}_{\{\tau_{2m-2} \leq t < \tau_{2m-1}\}} \left(\mathbb{E}\left[\widehat{\mathbb{1}}_{B_{2m-1}^{R+\kappa} A_{2m-1}^{T_i-D} B_{2m}^{T_i}} \,\Big|\, \mathcal{G}_t \right] \Bigg|_{\kappa=\tau_{2m-2}} + \sum_{j=m+1}^{\infty} \mathbb{E}\left[\widehat{\mathbb{1}}_{D_{2j-1}^R A_{2j-1}^{T_i-D} B_{2j}^{T_i}} \,\Big|\, \mathcal{G}_t \right] \right)$$

$$+ \sum_{m=2}^{\infty} \mathbb{1}_{\{\tau_{2m-1} \leq t < \tau_{2m}\}} \left(\mathbb{1}_{D_{2m-1}^R B_{2m-1}^W A_{2m-1}^{T_i-D}} \mathbb{E}\left[\widehat{\mathbb{1}}_{B_{2m}^{T_i}} \,\Big|\, \mathcal{G}_t \right] + \sum_{j=m+1}^{\infty} \mathbb{E}\left[\widehat{\mathbb{1}}_{D_{2j-1}^R A_{2j-1}^{T_i-D} B_{2j}^{T_i}} \,\Big|\, \mathcal{G}_t \right] \right) \Bigg\} . \tag{2.7}$$

Formulas (2.6) and (2.7) provide the basis for the two pricing frameworks which we will present in the sequel. It is apparent that we have to investigate the joint (conditional) distributions of the $\mathbb{P}$-numéraire portfolio S^{δ^*} and the different *sojourn-times* $\tau_j - \tau_{j-1}$, $j \geq 1$, in the respective states in order to calculate the fair insurance premium.

Equations (2.6) and (2.7) appear a bit nasty, not least because they contain infinite sums. These emerge due to the fact that a priori we have to consider all possible jumps between the states of unemployment and employment which may occur during the contract's duration. For simulations, however, we do not have to consider all jumps because the employment-unemployment process X is supposed to be right-continuous with finite state-space and hence càdlàg. It is a well known fact that each realization of such a process has only finitely many jumps with absolute jump size bigger than $\varepsilon > 0$ on every compact interval $[0, T]$. Therefore, every (simulated) path of X has only finitely many jumps on the interval $[0, T]$. We will exploit this feature within the second pricing framework which is based on Monte Carlo simulations and is presented in Section 2.3.

In the following section, we assume X to follow a time-homogeneous $\mathbb{F}$-doubly stochastic Markov chain. In this setting we get rid of the infinite sums, because they represent some well-known function.

2.2. The Insurance Premium for Time-Homogeneous $\mathbb{F}$-Doubly Stochastic Markov Chains

We now present a first framework in which we further specify the insurance premiums (2.6) and (2.7).

To this end, we assume the employment-unemployment process X to follow a time-homogeneous $\mathbb{F}$-doubly stochastic Markov chain as defined in Appendix A.2. In particular, the filtration $\mathbb{G}$ now satisfies $\mathbb{G} = \mathbb{F} \vee \mathbb{F}^X$ where $\mathbb{F}^X$ is the natural filtration generated by X and $\mathbb{F}$ is some arbitrary reference filtration satisfying the usual conditions with $\mathcal{F}_0 = \{\emptyset, \Omega\}$. In most cases of practical relevance, X is not taken to be adapted to $\mathbb{F}$ such that the jump times, defined in (2.1) are obviously $\mathbb{G}$- but not $\mathbb{F}$-stopping times.

In the following, we assume all portfolios $S^\delta \in \mathcal{V}_x$, introduced in Section 1.1, to be $\mathbb{F}$-adapted. In particular, the $\mathbb{P}$-numéraire portfolio S^{δ^*} is $\mathbb{F}$-adapted. Note that all assumptions in this more specific setting are compatible with the ones made in Chapter 1.

In order to model conditional distributions of the sojourn-times $(\tau_j - \tau_{j-1})_{j \geq 1}$, we assume X to admit an $\mathcal{F}_T$-measurable matrix-valued intensity $\Psi^* = [\psi_{j,k}^*]_{j,k \in \{1,2\}}$. More specifically, as we are interested in modeling the influences of risk-factors to the insurance premium, we assume X to be generated by a particular $\mathcal{F}_T$-measurable matrix Ψ^* which is possible due to Theorem A.7.

Assumption 2.2.1. *The employment-unemployment process X follows a time-homogeneous $\mathbb{F}$-doubly stochastic Markov chain with intensity matrix of the form*

$$\Psi^* = \begin{pmatrix} -\psi_1^* & \psi_1^* \\ \psi_2^* & -\psi_2^* \end{pmatrix} := \begin{pmatrix} -\psi_1(S_T^{\delta^*}) & \psi_1(S_T^{\delta^*}) \\ \psi_2(S_T^{\delta^*}) & -\psi_2(S_T^{\delta^*}) \end{pmatrix} , \tag{2.8}$$

where $\psi_{1,2} : (\mathbb{R}_+, \mathcal{B}(\mathbb{R}_+)) \mapsto (\mathbb{R}_+, \mathcal{B}(\mathbb{R}_+))$ are measurable functions.

According to Proposition A.19, an important property of the sojourn-times $\tau_j - \tau_{j-1}$, $j \geq 1$ of the time-homogeneous $\mathbb{F}$-doubly stochastic Markov chain X is then that

$$\mathbb{P}\left(\tau_j - \tau_{j-1} > t \middle| \mathcal{F}_T \vee \mathcal{F}^X_{\tau_{j-1}}\right) = e^{-\psi^*_{X_{\tau_{j-1}}} t} = e^{-t\psi^*_{1+((j+1) \bmod 2)}}, \quad t \in \mathbb{R}_+, \qquad (2.9)$$

provided $\tau_j < \infty$ $\mathbb{P}$-a.s.

Let us return to the Formulas (2.6) and (2.7). For the moment we assume that $t < W$ and that the first jump to unemployment τ_1 has not occurred up to time t ($t < \tau_1$). According to Equation (2.6), because of Remark A.17 2), and due to the time-homogeneity of X, we obtain

$$\pi_t(D) = \sum_{i=1}^{N} \mathbb{1}_{(t,T]}(T_i) c_i S_t^{\delta^*} \left\{ \mathbb{E}\left[\widehat{\mathbb{1}}_{\{W < \tau_1 \leq T_i - D, T_i < \tau_2\}} \middle| \mathcal{G}_t\right] \right.$$
$$\left. + \sum_{j=2}^{\infty} \mathbb{E}\left[\widehat{\mathbb{1}}_{\{\tau_{2j-1} - \tau_{2j-2} > R, W < \tau_{2j-1} \leq T_i - D, T_i < \tau_{2j}\}} \middle| \mathcal{G}_t\right] \right\}$$

$$= \sum_{i=1}^{N} \mathbb{1}_{(t,T]}(T_i) c_i S_t^{\delta^*} \left\{ \mathbb{E}\left[\mathbb{E}\left[\frac{1}{S_{T_i}^{\delta^*}}\mathbb{1}_{\{W < \tau_1 \leq T_i - D, T_i < \tau_2\}} \middle| \mathcal{F}_T \vee \mathcal{F}^X_t\right] \middle| \mathcal{G}_t\right] \right.$$
$$\left. + \sum_{j=2}^{\infty} \mathbb{E}\left[\mathbb{E}\left[\frac{1}{S_{T_i}^{\delta^*}}\mathbb{1}_{\{\tau_{2j-1} - \tau_{2j-2} > R, W < \tau_{2j-1} \leq T_i - D, T_i < \tau_{2j}\}} \middle| \mathcal{F}_T \vee \mathcal{F}^X_t\right] \middle| \mathcal{G}_t\right] \right\}$$

$$= \sum_{i=1}^{N} \mathbb{1}_{(t,T]}(T_i) c_i S_t^{\delta^*} \left\{ \mathbb{E}\left[\frac{1}{S_{T_i}^{\delta^*}}\mathbb{P}(W < \tau_1 \leq T_i - D, T_i < \tau_2 | \mathcal{F}_T \vee \mathcal{F}^X_t) \middle| \mathcal{G}_t\right] \right.$$
$$\left. + \sum_{j=2}^{\infty} \mathbb{E}\left[\frac{1}{S_{T_i}^{\delta^*}}\mathbb{P}(\tau_{2j-1} - \tau_{2j-2} > R, W < \tau_{2j-1} \leq T_i - D, T_i < \tau_{2j} | \mathcal{F}_T \vee \mathcal{F}^X_t) \middle| \mathcal{G}_t\right] \right\}$$

$$= \sum_{i=1}^{N} \mathbb{1}_{(t,T]}(T_i) c_i S_t^{\delta^*} \left\{ \mathbb{E}\left[\frac{1}{S_{T_i}}\mathbb{P}(W < \tau_1 \leq T_i - D, T_i < \tau_2 | \mathcal{F}_T \vee \sigma(X_t)) \middle| \mathcal{G}_t\right] \right.$$
$$\left. + \sum_{j=2}^{\infty} \mathbb{E}\left[\frac{1}{S_{T_i}^{\delta^*}}\mathbb{P}(\tau_{2j-1} - \tau_{2j-2} > R, W < \tau_{2j-1} \leq T_i - D, T_i < \tau_{2j} | \mathcal{F}_T \vee \sigma(X_t)) \middle| \mathcal{G}_t\right] \right\}$$

$$= \sum_{i=1}^{N} \mathbb{1}_{(t,T]}(T_i) c_i S_t^{\delta^*} \left\{ \mathbb{E}\left[\frac{1}{S_{T_i}^{\delta^*}}\underbrace{\mathbb{P}(W - t < \tau_1 \leq T_i - D - t, T_i - t < \tau_2 | \mathcal{F}_T)}_{(A)} \middle| \mathcal{G}_t\right] \right.$$
$$\left. + \sum_{j=2}^{\infty} \mathbb{E}\left[\frac{1}{S_{T_i}^{\delta^*}}\underbrace{\mathbb{P}(\tau_{2j-1} - \tau_{2j-2} > R, W - t < \tau_{2j-1} \leq T_i - D - t, T_i - t < \tau_{2j} | \mathcal{F}_T)}_{(B)} \middle| \mathcal{G}_t\right] \right\},$$
$$(2.10)$$

where we used the fact that $\sigma(X_0) = \{\emptyset, \Omega\}$.

Before we continue with the evaluations, we first define the notion of conditional (joint) densities, given the sigma algebra $\mathcal{F}_T$.

Definition 2.2.2. *We say that the $\mathbb{R}^n$-valued random vector $(X^1, ..., X^n)$, $n \geq 1$ on $(\Omega, \mathcal{G}, \mathbb{P})$ has the joint conditional density function $f_{X^1,...,X^n | \mathcal{F}_T} : (\Omega, \mathbb{R}^n) \to \mathbb{R}_+$ if the mapping $K : (\Omega, \mathcal{B}(\mathbb{R})^{\otimes n}) \to ([0,1], \mathcal{B}([0,1]))$, defined by*

$$K(\omega, A) := \int_A f_{X^1,...,X^n | \mathcal{F}_T}(\omega, s_1, ..., s_n) d\lambda^n(s_1, ..., s_n)$$

is a regular conditional distribution of $(X^1, ..., X^n)$ given $\mathcal{F}_T$. Here, λ^n denotes the Lebesgue measure on $(\mathbb{R}^n, \mathcal{B}(\mathbb{R})^{\otimes n})$.

Proposition 2.2.3. *Provided $\tau_m < \infty$ $\mathbb{P}$-a.s. for some $m \geq 1$, the family $(\tau_j - \tau_{j-1})_{1 \leq j \leq m}$ of sojourn-times is conditionally independent given $\mathcal{F}_T$ and for $1 \leq j \leq m$, the sojourn-time $\tau_j - \tau_{j-1}$ is conditionally exponential distributed to the parameter $\psi^*_{1+((j+1) \bmod 2)}$ given $\mathcal{F}_T$ in the sense that*

$$f_{\tau_j - \tau_{j-1} | \mathcal{F}_T}(\omega, t) := \psi^*_{1+((j+1) \bmod 2)}(\omega) e^{-t\psi^*_{1+((j+1) \bmod 2)}(\omega)}$$

is the conditional density function of $\tau_j - \tau_{j-1}$ given $\mathcal{F}_T$.

Proof. If $\tau_m < \infty$ $\mathbb{P}$-a.s. for some $m \geq 1$, then (2.9) implies that for all $1 \leq j \leq m$ we have

$$\mathbb{P}\left(\tau_j - \tau_{j-1} > t \middle| \mathcal{F}_T \vee \mathcal{F}^X_{\tau_{j-1}}\right) = e^{-t\psi^*_{1+((j+1) \bmod 2)}} = \mathbb{P}\left(\tau_j - \tau_{j-1} > t \middle| \mathcal{F}_T\right). \tag{2.11}$$

Because the open intervals (t, ∞), $t \geq 0$, generate the sigma algebra $\mathcal{B}(\mathbb{R}_+)$, this shows that the sojourn-time $\tau_j - \tau_{j-1}$ is conditionally independent of $\mathcal{F}^X_{\tau_{j-1}}$ given $\mathcal{F}_T$. In particular, the family $(\tau_j - \tau_{j-1})_{1 \leq j \leq m}$ of sojourn-times is conditionally independent given $\mathcal{F}_T$.

Consider now the mapping $K^j : \Omega \times \mathcal{B}(\mathbb{R}_+) \to [0,1]$ with

$$K^j(\omega, A) := \int_A \psi^*_{1+((j+1) \bmod 2)}(\omega) e^{-s\psi^*_{1+((j+1) \bmod 2)}(\omega)} ds.$$

We show that K^j is a regular conditional distribution of $\tau_j - \tau_{j-1}$ given $\mathcal{F}_T$.

- For fixed $A \in \mathcal{B}(\mathbb{R}_+)$ we have that $K^j(\cdot, A) : \Omega \to ([0,1], \mathcal{B}([0,1]))$ is $\mathcal{F}_T$-measurable because the intensities ψ^*_1 and ψ^*_2 are $\mathcal{F}_T$-measurable by definition and because the function $f_A : (\mathbb{R}_+, \mathcal{B}(\mathbb{R}_+)) \to ([0,1], \mathcal{B}([0,1]))$ with $f_A(\lambda) = \int_A \lambda e^{-s\lambda} ds$ is continuous.

- For fixed $\omega \in \Omega$, it is clear that $K^j(\omega, \cdot) : \mathcal{B}(\mathbb{R}_+) \to [0,1]$ defines a probability measure.

- For $t \geq 0$ we have

$$K^j(\omega, (t, \infty)) = \int_{t+}^{\infty} \psi^*_{1+((j+1) \bmod 2)}(\omega) e^{-s\psi^*_{1+((j-1) \bmod 2)}(\omega)} ds$$

$$= e^{-t\psi^*_{1+((j-1) \bmod 2)}(\omega)}$$

$$= \mathbb{P}(\tau_j - \tau_{j-1} > t | \mathcal{F}_T).$$

Because the open intervals (t, ∞), $t \geq 0$, generate the sigma algebra $\mathcal{B}(\mathbb{R}_+)$, this proves the assertion. $\qquad\square$

Remark 2.2.4. *The statement and the proof of Proposition 2.2.3 contain a small inaccuracy. In order to agree with the set-up in Section 1.1 and because the unemployment insurance contract matures at time T, we describe the $\mathbb{F}$-doubly stochastic Markov chain X in this section only up to the finite time-horizon T. According to (2.1), every jump which occurs after time T is then considered to be ∞. In other words the subset "provided $\tau_j < \infty$ $\mathbb{P}$-a.s." of Proposition 2.2.3 implies that $\tau_k \leq T$ $\mathbb{P}$-a.s. such that more specifically we should investigate regular conditional distributions on the time interval $[0, T]$.*

Yet, it is shown in Jakubowski and Niewęgłowski [61] that the definitions and statements on $\mathbb{F}$-doubly stochastic Markov chains as presented in Appendix A more generally hold for the time-index $\mathbb{R}_+$. The interested reader may then consider the distribution results of this section to hold for a general $\mathbb{F}$-doubly stochastic Markov chain $\widetilde{X} = (\widetilde{X}_t)_{t \in \mathbb{R}_+}$ which coincides with X on the interval $[0, T]$. As the results for the insurance premium remain unchanged in this case, we continue working with X instead of $\widetilde{X}$ for notational convenience.

For $j \geq 1$, let $W_j := \tau_{2j-1} - \tau_{2j-2}$, $Y_j := \sum_{l=1}^{2j-2}(\tau_l - \tau_{l-1})$ and $Z_j := \tau_{2j} - \tau_{2j-1}$. Provided $\tau_{2j} < \infty$ $\mathbb{P}$-a.s., it follows then from Proposition 2.2.3 by straightforward computations (e.g. by integral transformations) that W_j, Y_j and Z_j are conditionally independent and conditionally gamma-distributed random variables with $W_j \sim \mathrm{Exp}(\psi^*_1)|_{\mathcal{F}_T}$, $Y_j \sim \mathrm{Ga}(j-1, \psi^*_1)|_{\mathcal{F}_T} * \mathrm{Ga}(j-1, \psi^*_2)|_{\mathcal{F}_T}$ and $Z_j \sim \mathrm{Exp}(\psi^*_2)|_{\mathcal{F}_T}$ given $\mathcal{F}_T$. More explicitly, the conditional density functions of W_j, Y_j and Z_j given $\mathcal{F}_T$ are given as

$$f_{W_j}|_{\mathcal{F}_T}(\omega, t) = \psi^*_1(\omega) e^{-t\psi^*_1(\omega)} \mathbb{1}_{(0,\infty)}(t),$$

$$f_{Y_j}|_{\mathcal{F}_T}(\omega, t) = \int_0^t \frac{(\psi^*_1(\omega)\psi^*_2(\omega))^{j-1}}{(j-2)!(j-2)!} s^{j-2}(t-s)^{j-1} e^{-s\psi^*_1(\omega)} e^{-(t-s)\psi^*_2(\omega)} ds \,\mathbb{1}_{(0,\infty)}(t), \quad (2.12)$$

$$f_{Z_j}|_{\mathcal{F}_T}(\omega, t) = \psi^*_2(\omega) e^{-t\psi^*_2(\omega)} \mathbb{1}_{(0,\infty)}(t).$$

Straightforward integral transformation arguments then show that the joint conditional density $f_{\tau_1, \tau_2}|_{\mathcal{F}_T}$ of $(\tau_1, \tau_2) = (W_1, W_1 + Z_1)$ given $\mathcal{F}_T$ is of the form

$$f_{\tau_1, \tau_2}|_{\mathcal{F}_T}(u, v) = \psi^*_1 e^{-\psi^*_1 u} \mathbb{1}_{(0,\infty)}(u) \psi^*_1 e^{-\psi^*_2(v-u)} \mathbb{1}_{(0,\infty)}(v - u). \quad (2.13)$$

Furthermore note that because of $\tau_0 = 0$ every jump time τ_j can be written as a telescoping sum

$$\tau_j = \sum_{l=1}^{j}(\tau_l - \tau_{l-1})\,. \tag{2.14}$$

In particular, we obtain by integral transformation arguments that, provided $\tau_{2j} < \infty$ P-a.s. for some $j \geq 2$, the joint conditional density $f_{\tau_{2j-1}-\tau_{2j-2},\tau_{2j-1},\tau_{2j}}\big|_{\mathcal{F}_T}$ of $(\tau_{2j-1} - \tau_{2j-2}, \tau_{2j-1}, \tau_{2j}) = (W_j, W_j + Y_j, W_j + Y_j + Z_j)$ given $\mathcal{F}_T$ is of the form

$$
\begin{aligned}
&f_{\tau_{2j-1}-\tau_{2j-2},\tau_{2j-1},\tau_{2j}}\big|_{\mathcal{F}_T}(u,v,w)\\
=&f_{X_j}|_{\mathcal{F}_T}(u)\mathbf{1}_{(0,\infty)}(u)f_{Y_j}|_{\mathcal{F}_T}(v-u)\mathbf{1}_{(0,\infty)}(v-u)f_{Z_j}|_{\mathcal{F}_T}(w-v)\mathbf{1}_{(0,\infty)}(w-v)\,. \tag{2.15}
\end{aligned}
$$

For the moment, let $\tau_{2m} < \infty$ and $\tau_{2m+1} = \infty$ P-a.s. for some $m \geq 2$. By (2.13) and (2.15) we then obtain for the conditional probabilities (A) and (B) in (2.10) that

$$
\begin{aligned}
(A) &= \int_{W-t}^{T_i-D-t}\int_{T_i-t}^{\infty} f_{\tau_1,\tau_2}\big|_{\mathcal{F}_T}(u,v)dudv\\
&= \frac{\psi_1^*}{(\psi_1^* - \psi_2^*)}e^{-\psi_2^*(T_i-t)}\left(e^{-(\psi_1^*-\psi_2^*)(W-t)} - e^{-(\psi_1^*-\psi_2^*)(T_i-D-t)}\right) \tag{2.16}
\end{aligned}
$$

and

$$
\begin{aligned}
(B) &= \int_{R}^{\infty}\int_{W-t}^{T_i-D-t}\int_{T_i-t}^{\infty} f_{\tau_{2j-1}-\tau_{2j-2},\tau_{2j-1},\tau_{2j}}\big|_{\mathcal{F}_T}(u,v,w)dudvdw\\
&= \psi_1^*e^{-\psi_2^*(T_i-t)}\int_{(W-t)\vee R}^{T_i-D-t} e^{\psi_2^*v}\int_{R}^{v} e^{-\psi_1^*u}f_{Y_j}|_{\mathcal{F}_T}(v-u)dudv\,, \tag{2.17}
\end{aligned}
$$

and therefore

$$
\begin{aligned}
\pi_t(D) = \sum_{i=k^*}^{N}c_iS_t^{\delta^*}&\left\{\mathbb{E}\left[\frac{\psi_1^*}{S_{T_i}^{\delta^*}(\psi_1^*-\psi_2^*)}e^{-\psi_2^*(T_i-t)}\left(e^{-(\psi_1^*-\psi_2^*)(W-t)} - e^{-(\psi_1^*-\psi_2^*)(T_i-D-t)}\right)\bigg|\mathcal{G}_t\right]\right.\\
&\left.+\sum_{j=2}^{m}\mathbb{E}\left[\frac{1}{S_{T_i}^{\delta^*}}\psi_1^*e^{-\psi_2^*(T_i-t)}\int_{(W-t)\vee R}^{T_i-D-t}e^{\psi_2^*y}\int_{R}^{y}e^{-\psi_1^*x}f_{Y_j}|_{\mathcal{F}_T}(y-x)dxdy\bigg|\mathcal{G}_t\right]\right\}.
\end{aligned}
$$

Here, k^* is the first $i \in \{1,...,N\}$ for which $T_i - D \geq W$ such that the integrals are all well defined. Note that because we are in the setting where $t < W$, we can suppress the indicator function $\mathbf{1}_{(t,T]}(T_i)$ of equation (2.6).

Remark 2.2.5. *The assumption $\tau_{2m} < \infty$ and $\tau_{2m+1} = \infty$ $\mathbb{P}$-a.s. for some $m \geq 2$ simplifies the evaluation of the insurance premium as we need to investigate the jump times only up to a finite number. Yet, the assumption appears artificial for practical examples as X is the individual employment-unemployment process of an insured person. For the remaining investigations of this section we rather make the following assumption.*

Assumption 2.2.6. *We assume $\tau_j < \infty$ $\mathbb{P}$-a.s. for all $j \geq 1$.*

With Assumption 2.2.6 as well as (2.16) and (2.17), we then obtain

$$
\pi_t(D) = \sum_{i=k^*}^{N} c_i S_t^{\delta^*} \left\{ \mathbb{E}\left[\frac{\psi_1^*}{S_{T_i}^{\delta^*}(\psi_1^* - \psi_2^*)} e^{-\psi_2^*(T_i - t)} \left(e^{-(\psi_1^* - \psi_2^*)(W-t)} - e^{-(\psi_1^* - \psi_2^*)(T_i - D - t)} \right) \,\middle|\, \mathcal{G}_t \right] \right.
$$
$$
\left. + \sum_{j=2}^{\infty} \mathbb{E}\left[\frac{1}{S_{T_i}^{\delta^*}} \psi_1^* e^{-\psi_2^*(T_i - t)} \int_{\max\{W-t, R\}}^{T_i - D - t} e^{\psi_2^* y} \int_{R}^{y} e^{-\psi_1^* x} f_{Y_j}|_{\mathcal{F}_T}(y - x)\,dx\,dy \,\middle|\, \mathcal{G}_t \right] \right\}.
$$

By monotone convergence, this yields

$$
\pi_t(D) = \sum_{i=k^*}^{N} c_i S_t^{\delta^*} \left\{ \mathbb{E}\left[\frac{\phi_1(S_T^{\delta^*})}{S_{T_i}^{\delta^*}} \,\middle|\, \mathcal{G}_t \right] + \mathbb{E}\left[\frac{\phi_2(S_T^{\delta^*})}{S_{T_i}^{\delta^*}} \,\middle|\, \mathcal{G}_t \right] \right\}, \tag{2.18}
$$

where

$$
\phi_1(S_T^{\delta^*}) := \frac{\psi_1^*}{(\psi_1^* - \psi_2^*)} e^{-\psi_2^*(T_i - t)} \left(e^{-(\psi_1^* - \psi_2^*)(W-t)} - e^{-(\psi_1^* - \psi_2^*)(T_i - D - t)} \right), \tag{2.19}
$$

$$
\phi_2(S_T^{\delta^*}) := (\psi_1^*)^2 \psi_2^* e^{-\psi_2^*(T_i - t)} \int_{(W-t)\vee R}^{T_i - D - t} \int_{R}^{v} e^{-(\psi_1^* - \psi_2^*)u} \int_{0}^{v-u} e^{-(\psi_1^* - \psi_2^*)x}
$$
$$
\times\ I_0\left(2\sqrt{\psi_1^* \psi_2^* x(v - u - x)} \right) dx\,du\,dv. \tag{2.20}
$$

Here, I_0 is the modified first kind Bessel function of order 0. In general, the modified first order Bessel function I_α of order $\alpha \in \mathbb{R}$ is given by

$$
I_\alpha(y) = \sum_{m=0}^{\infty} \frac{1}{m!\Gamma(m + \alpha + 1)} \left(\frac{y}{2} \right)^{2m+\alpha}. \tag{2.21}
$$

Recall that ψ_1^* and ψ_2^* and therefore also $\frac{\phi_1(S_T^{\delta^*})}{S_{T_i}^{\delta^*}}$ and $\frac{\phi_2(S_T^{\delta^*})}{S_{T_i}^{\delta^*}}$ are $\mathcal{F}_T$-measurable and non-negative. Since Hypothesis (H) holds[1] according to Proposition A.4 and Remark A.17 3), we finally have

---

[1]Note that $\frac{\phi_1(S_T^{\delta^*})}{S_{T_i}^{\delta^*}}$ and $\frac{\phi_2(S_T^{\delta^*})}{S_{T_i}^{\delta^*}}$ do not need to be bounded. However, one can truncate by n and use monotone convergence.

$$\pi_t(D) = \sum_{i=k^*}^{N} c_i S_t^{\delta^*} \left\{ \mathbb{E}\left[\frac{\phi_1(S_T^{\delta^*})}{S_{T_i}^{\delta^*}} \,\bigg|\, \mathcal{F}_t \right] + \mathbb{E}\left[\frac{\phi_2(S_T^{\delta^*})}{S_{T_i}^{\delta^*}} \,\bigg|\, \mathcal{F}_t \right] \right\}. \tag{2.22}$$

In the following proposition, we give the insurance premium $\pi_t(D)$ for the most important cases where $t < \tau_1$. Because the evaluations for the case $t \geq W$ are very similar to the ones for $t < W$, its proof is omitted.

Note that due to the "loss of memory" property of X, the results for the case $t < \tau_1$ are sufficient to obtain the insurance premiums also for all other cases appearing in Equations (2.6) and (2.7). We show this in more detail in Proposition 2.2.8 for the exemplary case where $t < W$ and $\tau_{2m-2} \leq t < \tau_{2m-1}$.

Proposition 2.2.7. *With Assumption 2.2.6 and for $t < \tau_1$, we obtain the insurance premiums $\pi_t(D)$ at time $t \in [0, T]$ after settlement of all payments up to and including time t as follows.*

- *If $t < W$, $\pi_t(D)$ is given in (2.22).*

- *If $t \geq W$, $\pi_t(D)$ is given as*

$$\pi_t(D) = \sum_{i=k^*}^{N} \mathbb{1}_{(t,T]}(T_i) c_i S_t^{\delta^*} \left\{ \mathbb{E}\left[\frac{\phi_3(S_T^{\delta^*})}{S_{T_i}^{\delta^*}} \,\bigg|\, \mathcal{F}_t \right] + \mathbb{E}\left[\frac{\phi_4(S_T^{\delta^*})}{S_{T_i}^{\delta^*}} \,\bigg|\, \mathcal{F}_t \right] \right\}, \tag{2.23}$$

where

$$\phi_3(S_T^{\delta^*}) := \frac{\psi_1^*}{(\psi_1^* - \psi_2^*)} e^{-\psi_2^*(T_i - t)} \left(1 - e^{-(\psi_1^* - \psi_2^*)(T_i - D - t)} \right)$$

$$\phi_4(S_T^{\delta^*}) := (\psi_1^*)^2 \psi_2^* e^{-\psi_2^*(T_i - t)} \int\limits_{R}^{T_i - D - t} \int\limits_{R}^{v} e^{-(\psi_1^* - \psi_2^*)u} \int\limits_{0}^{v-u} e^{-(\psi_1^* - \psi_2^*)x}$$

$$\times \, I_0\left(2\sqrt{\psi_1^* \psi_2^* x (v - u - x)} \right) dx\,du\,dv.$$

Proposition 2.2.8. *With Assumption 2.2.6 and for $t < W$, let $\tau_{2m-2} \leq t < \tau_{2m-1}$ for some $m \geq 2$. Then the insurance premium $\pi_t(D)$ at time $t \in [0, W)$ is given as*

$$\pi_t(D) = \sum_{i=k^*}^{N} c_i S_t^{\delta^*} \left\{ \mathbb{E}\left[\frac{\phi_5(S_T^{\delta^*})}{S_{T_i}^{\delta^*}} \,\bigg|\, \mathcal{F}_t \right] + \mathbb{E}\left[\frac{\phi_6(S_T^{\delta^*})}{S_{T_i}^{\delta^*}} \,\bigg|\, \mathcal{F}_t \right] \right\}, \tag{2.24}$$

where

$$\phi_5(S_T^{\delta^*}) := \frac{\psi_1^*}{(\psi_1^* - \psi_2^*)} e^{-\psi_2^*(T_i-t)} \left(e^{-(\psi_1^*-\psi_2^*)(\max\{W,R+\tau_{2m-2}\}-t)} - e^{-(\psi_1^*-\psi_2^*)(T_i-D-t)} \right)$$

$$\phi_6(S_T^{\delta^*}) := (\psi_1^*)^2 \psi_2^* e^{-\psi_2^*(T_i-t)} \int\limits_{\max\{W-t,R\}}^{T_i-D-t} \int\limits_{R}^{y} e^{-(\psi_1^*-\psi_2^*)x} \int\limits_{0}^{y-x} e^{-(\psi_1^*-\psi_2^*)u}$$

$$\times \ I_0(2\sqrt{\psi_1^*\psi_2^* u(y-x-u)}) du dx dy \,.$$

Proof. By equation (2.6) we have

$$\pi_t(D) = \sum_{i=k^*}^{N} c_i S_t^{\delta^*} \left\{ \mathbb{E}\left[\widehat{\mathbb{1}}_{\{W\vee(R+\kappa)<\tau_{2m-1}\leq T_i-D,\tau_{2m}>T_i\}} \,\Big|\, \mathcal{G}_t \right]\Big|_{\kappa=\tau_{2m-2}} \right.$$

$$\left. + \sum_{j=m+1}^{\infty} \mathbb{E}\left[\widehat{\mathbb{1}}_{\{\tau_{2j-1}-\tau_{2j-2}>R,W<\tau_{2j-1}\leq T_i-D,\tau_{2j}>T_i\}} \,\Big|\, \mathcal{G}_t \right] \right\}$$

$$= \sum_{i=k^*}^{N} c_i S_t^{\delta^*} \left\{ \mathbb{E}\left[\mathbb{E}\left[\frac{1}{S_{T_i}^{\delta^*}} \mathbb{1}_{\{W\vee(R+\kappa)<\tau_{2m-1}\leq T_i-D,\tau_{2m}>T_i\}} \,\Big|\, \mathcal{F}_T\vee\mathcal{F}_t^X \right] \Big|\, \mathcal{G}_t \right]\Big|_{\kappa=\tau_{2m-2}} \right.$$

$$\left. + \sum_{j=m+1}^{\infty} \mathbb{E}\left[\mathbb{E}\left[\frac{1}{S_{T_i}^{\delta^*}} \mathbb{1}_{\{\tau_{2j-1}-\tau_{2j-2}>R,W<\tau_{2j-1}\leq T_i-D,\tau_{2j}>T_i\}} \,\Big|\, \mathcal{F}_T\vee\mathcal{F}_t^X \right] \Big|\, \mathcal{G}_t \right] \right\}$$

$$= \sum_{i=k^*}^{N} c_i S_t^{\delta^*} \left\{ \mathbb{E}\left[\frac{1}{S_{T_i}^{\delta^*}} \mathbb{P}\left(W\vee(R+\kappa)<\tau_{2m-1}\leq T_i-D,\tau_{2m}>T_i \big| \mathcal{F}_T\vee\mathcal{F}_t^X \right) \Big|\, \mathcal{G}_t \right]\Big|_{\kappa=\tau_{2m-2}} \right.$$

$$\left. + \sum_{j=m+1}^{\infty} \mathbb{E}\left[\frac{1}{S_{T_i}^{\delta^*}} \mathbb{P}\left(\tau_{2j-1}-\tau_{2j-2}>R,W<\tau_{2j-1}\leq T_i-D,\tau_{2j}>T_i \big| \mathcal{F}_T\vee\mathcal{F}_t^X \right) \Big|\, \mathcal{G}_t \right] \right\}$$

$$= \sum_{i=k^*}^{N} c_i S_t^{\delta^*} \left\{ \mathbb{E}\left[\frac{1}{S_{T_i}^{\delta^*}} \mathbb{P}\left(W\vee(R+\kappa)<\tau_{2m-1}\leq T_i-D,\tau_{2m}>T_i \big| \mathcal{F}_T\vee\sigma(X_t) \right) \Big|\, \mathcal{G}_t \right]\Big|_{\kappa=\tau_{2m-2}} \right.$$

$$\left. + \sum_{j=m+1}^{\infty} \mathbb{E}\left[\frac{1}{S_{T_i}^{\delta^*}} \mathbb{P}\left(\tau_{2j-1}-\tau_{2j-2}>R,W<\tau_{2j-1}\leq T_i-D,\tau_{2j}>T_i \big| \mathcal{F}_T\vee\sigma(X_t) \right) \Big|\, \mathcal{G}_t \right] \right\}$$

$$= \sum_{i=k^*}^{N} c_i S_t^{\delta^*} \left\{ \mathbb{E}\left[\frac{1}{S_{T_i}^{\delta^*}} \mathbb{P}\left(W\vee(R+\kappa)-t<\tau_{1}\leq T_i-D-t,\tau_{2}>T_i-t \big| \mathcal{F}_T \right) \Big|\, \mathcal{G}_t \right]\Big|_{\kappa=\tau_{2m-2}} \right.$$

$$\left. + \sum_{j=2}^{\infty} \mathbb{E}\left[\frac{1}{S_{T_i}^{\delta^*}} \mathbb{P}\left(\tau_{2j-1}-\tau_{2j-2}>R,W-t<\tau_{2j-1}\leq T_i-D-t,\tau_{2j}>T_i-t \big| \mathcal{F}_T \right) \Big|\, \mathcal{G}_t \right] \right\}.$$

Note that τ_{2m-1} is the first jump time after time t. Hence, by exploiting the time-homogeneity of X and executing a "time-reduction" by t, τ_{2m-1} becomes the first jump time of the "renewed" F-doubly stochastic Markov chain and the last equality holds. The result then follows immediately by applying the same arguments which we applied to (2.10) in order to obtain (2.18).   □

With the assumption $\psi_{1,2}^* = \psi_{1,2}(S_T^{\delta^*})$ of (2.8), we hence find insurance premiums which can be adjusted according to the conditional distribution of the P-numéraire portfolio S^{δ^*}. As indicated in Section 1.2, the P-numéraire portfolio is a global indicator of the hybrid market and hence a reasonable risk-factor for hybrid insurance products.

A further evaluation of the conditional expectations e.g. in Equations (2.22), (2.23) or (2.24) can now be handled in different ways. Analytic solutions like inverse Fourier methods may face some difficulties because the functions in the conditional expectations are rather hard to invert as they depend on the value of $S_t^{\delta^*}$ at the three different times t, T_i, and T. Alternatively, one could assume some dynamics for $S_t^{\delta^*}$ and then run Monte Carlo simulations. In the following subsection we provide a general example how to further specify $\pi_t(D)$ if we assume the P-numéraire portfolio to be a Lévy process.

2.2.1. The P-Numéraire Portfolio as a Lévy Process

In order to further specify the insurance premiums, we now let the P-numéraire portfolio S^{δ^*} be a Lévy process with Lévy Khintchine characteristics (a, b, ν) and such that its distribution at time $t \in [0, T]$ provides a density h_t^* with respect to the Lebesgue measure given by

$$h_t^*(x) = \frac{1}{2\pi} \int_{\mathbb{R}} e^{-iyx} \varphi_{S_t^{\delta^*}}(y) dy \,, \tag{2.25}$$

where $\varphi_{S_t^{\delta^*}}$ is the characteristic function of $S_t^{\delta^*}$ at time $t \in [0, T]$, equal to

$$\varphi_{S_t^{\delta^*}}(y) = \exp\left(-t\left(\frac{b^2}{2} y^2 - iay + \int_{\{|x| \geq 1\}} (1 - e^{iyx})\nu(dx) + \int_{\{|x| < 1\}} (1 - e^{iyx} + iyx)\nu(dx) \right) \right) . \tag{2.26}$$

Moreover, we assume $\mathbb{F} = \mathbb{F}^{S^{\delta^*}}$.

Based on these assumptions and given Assumption 2.2.6, we can now further specify the insurance premium $\pi_t(D)$. We show the exemplary calculations only for the most important case $t < W$ and $t < \tau_1$. With the notations (2.19) and (2.20), we then obtain

$$\pi_t(D) = \sum_{i=k^*}^N c_i S_t^{\delta^*} \left\{ \mathbb{E}\left[\frac{\phi_1(S_T^{\delta^*})}{S_{T_i}^{\delta^*}} \,\Big|\, \mathcal{F}_t \right] + \mathbb{E}\left[\frac{\phi_2(S_T^{\delta^*})}{S_{T_i}^{\delta^*}} \,\Big|\, \mathcal{F}_t \right] \right\}$$

$$= \sum_{i=k^*}^N c_i S_t^{\delta^*} \left\{ \mathbb{E}\left[\mathbb{E}\left[\frac{\phi_1(S_T^{\delta^*} - S_{T_i}^{\delta^*} + S_{T_i}^{\delta^*})}{S_{T_i}^{\delta^*}} \,\Big|\, \mathcal{F}_{T_i} \right] \,\Big|\, \mathcal{F}_t \right] \right.$$

$$+ \mathbb{E}\left[\mathbb{E}\left[\left.\frac{\phi_2(S_T^{\delta^*} - S_{T_i}^{\delta^*} + S_{T_i}^{\delta^*})}{S_{T_i}^{\delta^*}}\,\right|\, \mathcal{F}_{T_i}\right]\,\bigg|\, \mathcal{F}_t\right]\Bigg\}$$

$$= \sum_{i=k^*}^{N} c_i S_t^{\delta^*}\left\{\mathbb{E}\left[\left.\mathbb{E}\left[\left.\frac{\phi_1(S_T^{\delta^*} - S_{T_i}^{\delta^*} + x)}{x}\right]\right|_{x=S_{T_i}^{\delta^*}}\,\right|\, \mathcal{F}_t\right]\right.$$

$$\left.+ \mathbb{E}\left[\left.\mathbb{E}\left[\left.\frac{\phi_2(S_T^{\delta^*} - S_{T_i}^{\delta^*} + x)}{x}\right]\right|_{x=S_{T_i}^{\delta^*}}\,\right|\, \mathcal{F}_t\right]\right\}$$

$$= \sum_{i=k^*}^{N} c_i S_t^{\delta^*}\left\{\mathbb{E}\left[\left.\int_{\mathbb{R}}\frac{\phi_1(y+x)}{x}h_{T-T_i}^*(y)dy\,\right|_{x=S_{T_i}^{\delta^*}}\,\bigg|\, \mathcal{F}_t\right]\right.$$

$$\left.+ \mathbb{E}\left[\left.\int_{\mathbb{R}}\frac{\phi_2(y+x)}{x}h_{T-T_i}^*(y)dy\,\right|_{x=S_{T_i}^{\delta^*}}\,\bigg|\, \mathcal{F}_t\right]\right\}$$

$$= \sum_{i=k^*}^{N} c_i S_t^{\delta^*}\left\{\mathbb{E}\left[\underbrace{\int_{\mathbb{R}}\frac{\phi_1(y+S_{T_i}^{\delta^*})}{S_{T_i}^{\delta^*}}h_{T-T_i}^*(y)dy}_{=:\lambda_1(S_{T_i}^{\delta^*})}\,\bigg|\, \mathcal{F}_t\right]\right.$$

$$\left.+ \mathbb{E}\left[\underbrace{\int_{\mathbb{R}}\frac{\phi_2(y+S_{T_i}^{\delta^*})}{S_{T_i}^{\delta^*}}h_{T-T_i}^*(y)dy}_{=:\lambda_2(S_{T_i}^{\delta^*})}\,\bigg|\, \mathcal{F}_t\right]\right\}$$

$$= \sum_{i=k^*}^{N} c_i S_t^{\delta^*}\left\{\mathbb{E}\left[\lambda_1(S_{T_i}^{\delta^*} - S_t^{\delta^*} + S_t^{\delta^*})\,\big|\, \mathcal{F}_t\right] + \mathbb{E}\left[\lambda_2(S_{T_i}^{\delta^*} - S_t^{\delta^*} + S_t^{\delta^*})\,\big|\, \mathcal{F}_t\right]\right\}$$

$$= \sum_{i=k^*}^{N} c_i S_t^{\delta^*}\left\{\mathbb{E}\left[\lambda_1(S_{T_i}^{\delta^*} - S_t^{\delta^*} + x)\right]\Big|_{x=S_t^{\delta^*}} + \mathbb{E}\left[\lambda_2(S_{T_i}^{\delta^*} - S_t^{\delta^*} + x)\right]\Big|_{x=S_t^{\delta^*}}\right\}$$

$$= \sum_{i=k^*}^{N} c_i S_t^{\delta^*}\left\{\int_{\mathbb{R}}\lambda_1(z+x)h_{T_i-t}^*(z)dz\,\bigg|_{x=S_t^{\delta^*}} + \int_{\mathbb{R}}\lambda_2(z+x)h_{T_i-t}^*(z)dz\,\bigg|_{x=S_t^{\delta^*}}\right\}$$

$$= \sum_{i=k^*}^{N} c_i S_t^{\delta^*}\left\{\int_{\mathbb{R}}\lambda_1(z+S_t^{\delta^*})h_{T_i-t}^*(z)dz + \int_{\mathbb{R}}\lambda_2(z+S_t^{\delta^*})h_{T_i-t}^*(z)dz\right\}$$

$$= \sum_{i=k^*}^{N} c_i S_t^{\delta^*}\left\{\int_{\mathbb{R}}\int_{\mathbb{R}}\frac{\phi_1(y+z+S_t^{\delta^*})}{z+S_t^{\delta^*}}h_{T-T_i}^*(y)dy\,h_{T_i-t}^*(z)dz\right.$$

$$\left.+ \int_{\mathbb{R}}\int_{\mathbb{R}}\frac{\phi_2(y+z+S_t^{\delta^*})}{z+S_t^{\delta^*}}h_{T-T_i}^*(y)dy\,h_{T_i-t}^*(z)dz\right\}$$

$$= \sum_{i=k^*}^{N} c_i S_t^{\delta^*} \left\{ \iint_{\mathbb{R}} \frac{\psi_1(y + z + S_t^{\delta^*})}{(z + S_t^{\delta^*})(\psi_1(y + z + S_t^{\delta^*}) - \psi_2(y + z + S_t^{\delta^*}))} e^{-\psi_2(y+z+S_t^{\delta^*})(T_i - t)} \right.$$

$$\times \left(e^{-(\psi_1(y+z+S_t^{\delta^*}) - \psi_2(y+z+S_t^{\delta^*}))(W-t)} - e^{-(\psi_1(y+z+S_t^{\delta^*}) - \psi_2(y+z+S_t^{\delta^*}))(T_i - D - t)} \right) h_{T-T_i}^*(y)$$

$$\times h_{T_i-t}^*(z) dy dz \; + \; \iint_{\mathbb{R}} \left[\frac{(\psi_1(y + z + S_t^{\delta^*}))^2 \psi_2(y + z + S_t^{\delta^*}) e^{-\psi_2(y+z+S_t^{\delta^*})(T_i - t)}}{z + S_t^{\delta^*}} \right.$$

$$\times \int_{(W-t)\vee\mathbb{R}}^{T_i - D - t} \int_{\mathbb{R}}^{v} e^{-(\psi_1(y+z+S_t^{\delta^*}) - \psi_2(y+z+S_t^{\delta^*}))u} \int_0^{v-u} e^{-(\psi_1(y+z+S_t^{\delta^*}) - \psi_2(y+z+S_t^{\delta^*}))x}$$

$$\left. \times I_0 \left(2\sqrt{\psi_1(y + z + S_t^{\delta^*})\psi_2(y + z + S_t^{\delta^*})x(v - u - x)} \right) dx du dv \right]$$

$$\left. \times h_{T-T_i}^*(y) h_{T_i-t}^*(z) dy dz \right\}, \tag{2.27}$$

where h^* is given by (2.25) and (2.26) and ψ_1, ψ_2 are the measurable functions, defined in (2.8).

In this particular Lévy process set-up for the $\mathbb{P}$-numéraire portfolio we are therefore able to compute the insurance premium $\pi_t(D)$ given the value $S_t^{\delta^*}$ at time $t \in [0, T]$. In particular, (2.27) provides a good framework for simulations.

2.2.2. A Classical Markov Chain Setting

In this subsection we provide a sensitivity analysis for the insurance premium $\pi_0(D)$ at time $t = 0$ and an estimation procedure for the intensities ψ_1^*, ψ_2^* in a classical Markov chain setting. To this end, we assume $\mathcal{F}_t = \{\emptyset, \Omega\}$ for all $t \in [0, T]$ and hence that the $\mathbb{P}$-numéraire portfolio is a deterministic function in time. For the sake of simplicity, we assume $S_t^{\delta^*} = e^{rt}$, $t \in [0, T]$, for some constant rate $r \geq 0$.

In this setting, the expressions in the conditional expectation of formula (2.22) are deterministic and we obtain the premium $\pi_0(D)$ as

$$\pi_0(D) = \sum_{i=1}^{N} c_i e^{-rT_i} \left(\frac{\psi_1^*}{(\psi_1^* - \psi_2^*)} e^{-\psi_2^*(T_i - t)} \left(e^{-(\psi_1^* - \psi_2^*)(W-t)} - e^{-(\psi_1^* - \psi_2^*)(T_i - D - t)} \right) \right.$$

$$\left. + (\psi_1^*)^2 \psi_2^* e^{-\psi_2^*(T_i - t)} \int_{(W-t)\vee\mathbb{R}}^{T_i - D - t} \int_{\mathbb{R}}^{v} e^{-(\psi_1^* - \psi_2^*)u} \int_0^{v-u} e^{-(\psi_1^* - \psi_2^*)x} I_0 \left(2\sqrt{\psi_1^* \psi_2^* x(v - u - x)} \right) dx du dv \right),$$

$$\tag{2.28}$$

for $\psi_1^*, \psi_2^* \in \mathbb{R}_+$.

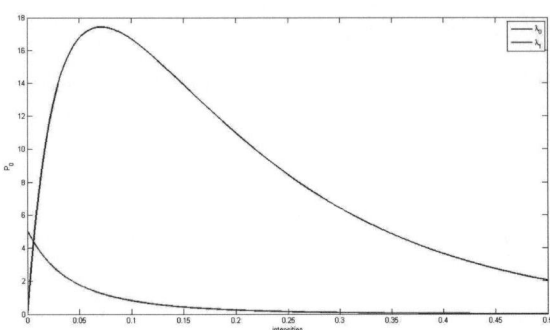

Figure 2.3.: Sensitivity of the insurance premium $\pi_0(D)$ to the intensity parameters ψ_1^* and ψ_2^* in the classical time-homogeneous Markov chain setting.

Sensitivity Analysis

In order to test the model on its accuracy, we investigate (2.28) on its sensitivity towards changes in the intensities ψ_1^* and ψ_2^* as well as towards the time periods W, D, and R. To this end, we let the respective parameter of interest vary and fix the other parameters of formula (2.28) to one of the following levels:

$$T_N = 72 \text{(months)} , \ c_i = 1 , \ \psi_1^* = 0,36\% , \ \psi_2^* = 2,7\% ,$$
$$W = 6 \text{ (months)} , \ D = 6 \text{ (months)} , \ R = 6 \text{ (months)} , \ r = 4\%$$

- **The intensities ψ_1^* and ψ_2^*:**
  Figure 2.3 shows the characteristics of the insurance premium due to changes in the intensities ψ_1^* and ψ_2^*. The results cover the anticipated behaviour of the insurance premium: because of the exponential distributions in this setting, we have that $\mathbb{E}[\tau_{2j-1} - \tau_{2j-2}] = \frac{1}{\psi_1^*}$ and $\mathbb{E}[\tau_{2j} - \tau_{2j-1}] = \frac{1}{\psi_2^*}$ for $j \geq 1$. Therefore, an increase in ψ_1^* is equal to a decrease in the expected time of employment. As a consequence, the insurance premium has first to increase. In the (unrealistic) case where the expected time of employment reaches the low level of 12 months ($\psi_1^* \approx 0,083$), more and more claims are not paid because the employment time does often not exceed the waiting plus deferment or the requalification plus deferment period. Consequently, the insurance premium decreases.

  Analogously, an increase in ψ_2^* is equal to a decrease in the expected time of unemployment and the insurance premium has to decrease.

- **Waiting, deferment and requalification period:**
  Figure 2.4 shows the changes in the insurance premium due to variations in the

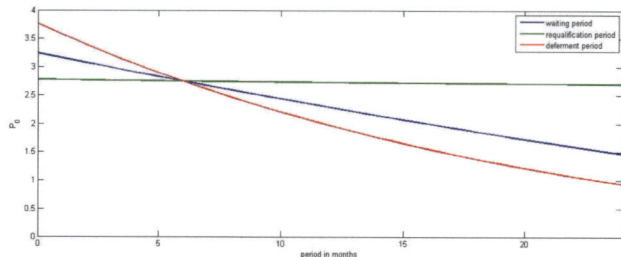

Figure 2.4.: Sensitivity of the insurance premium $\pi_0(D)$ to the waiting, deferment and requalification periods in the classical time-homogeneous Markov chain setting.

waiting, deferment, and requalification periods. Again, the results cover the anticipated behaviour of the insurance premium: an increase in all three periods decreases the probability of claim payments and should therefore result in lower insurance premiums.

Note that the insurance premium reacts more sensitively to changes in the waiting and the deferment period than to changes in the requalification period.

Estimation of the Intensities

The insurance premium, given in Equation (2.28), is a function of several parameters. Most of them are fixed in the terms of contract like the waiting, deferment, and requalification period, the maturity, the payments c_i, or their respective dates T_i. A priori unknown parameters are the intensities ψ_1^* and ψ_2^* as well as the parameter r. We now present estimation results for the intensities ψ_1^* and ψ_2^* which are based on real data published by the German federal employment office[2].

Because X in this simple setting is supposed to follow a time-homogeneous Markov chain, we can apply the estimation methods presented in Kalbfleisch and Lawless [64]. Here, a constant quantity of persons is observed over a certain period. Each person is supposed to be in a certain observable state at any time of the period, where the set of states is finite (in our case the states are: "employed" and "unemployed"). At some fixed time points of the period the respective state of each person in the group is listed.

For every person, the random process of jumping from one state to another is supposed to follow a time-homogeneous Markov chain and all processes are supposed to be i.i.d. Under these assumptions, it is then possible to apply a maximum-likelihood estimation method to derive estimators for the components of the intensity matrix $\mathbf{\Psi}^*$.

In order to perform the estimations, we need

- the number n_{11}^i of persons that are in state 1 at time t_{i-1} and at time t_i,

---

[2]Source: http://www.arbeitsagentur.de/ .

- the number n_{12}^i of persons that moved from state 1 at time t_{i-1} to state 2 at time t_i,

- the number n_{22}^i of persons that are in state 2 at time t_{i-1} and at time t_i,

- the number n_{21}^i of persons that moved from state 2 at time t_{i-1} to state 1 at time t_i, $i = 1, ..., n$,

for a sequence $t_0, ..., t_n$ of points in time. Unfortunately, there is no public data available, containing the exact numbers. Therefore, we proceed in the following way:

We set the time lag $t_i - t_{i-1}$ equal to one month and extract approximations to the required numbers out of the monthly job market reports of the Federal Employment Office of Germany.

To this end, we interpret the labor force in Germany to consist of all employed and unemployed persons. The total number of unemployed persons as well as the unemployment rate are given in the reports. The unemployment rate however is the fraction of all unemployed persons with respect to the labor force. By using this information we can hence derive the total numbers of unemployed and employed persons, respectively, every month. These values are interpreted to be the number of persons that stay in the state "unemployed" or "employed", respectively. The monthly changes of the total values are interpreted to form the number of persons that are employed at time t_{i-1} and become unemployed until t_i and vice versa. Moreover, since the labor force of Germany is identified once a year (at the beginning of May), the observed samples remain constant over a time period of one year. This is important for our estimation method as the sample of observed people must remain constant over the observation period.

The estimation results, which base on monthly data from May to April $'98/'99$ to $'07/'08$ are presented in Table 2.1.

Note that the way of extracting the numbers contains some inaccuracies. First of all, the labor force is not constant over a time period of one year and may not only consist of employed and unemployed persons. Moreover, the numbers of changes from one state to another represent cumulated values and not the total values. Nevertheless, the presented results already provide first hints to the real dimensions of the intensities on the German labor market. In the next Section, the focus is on elaborating estimates for the intensities on the German labor market more thoroughly.

2.3. The Insurance Premium for General $\mathbb{F}$-Doubly Stochastic Markov Chains

We have seen in Section 2.2, that the general insurance premiums (2.6) and (2.7) can be further specified in a time-homogeneous $\mathbb{F}$-doubly stochastic setting to an insurance premium which only depends on information contained in the reference filtration $\mathbb{F}$. More specifically, in Section 2.2.1 we have seen that if the filtration is generated by the $\mathbb{P}$-numéraire portfolio S^{δ^*} given as a Lévy process, than any path of the fair price process

	$\widehat{\psi_1^*}$	$\widehat{\psi_2^*}$
'98/'99	0,0038552007	0,0282108984
'99/'00	0,0036361118	0,0281011314
'00/'01	0,0034410489	0,0284704359
'01/'02	0,0036146433	0,0265663626
'02/'03	0,0040511845	0,0251325837
'03/'04	0,0042670366	0,0263048480
'04/'05	0,0041309233	0,0244147955
'05/'06	0,0035819277	0,0243732156
'06/'07	0,0027738128	0,0247146137
'07/'08	0,0030655450	0,0325999498

Table 2.1.: Maximum-Likelihood estimators $\widehat{\psi_1^*}$ and $\widehat{\psi_2^*}$ of the intensities ψ_1^* and ψ_2^* in the time-homogeneous (classical) Markov chain setting for the periods $'98/'99$ - $'07/'08$.

$(\pi_t(D))_{t\in[0,T]}$ is given provided one "knows" the path of S^{δ^*}. The $\mathbb{P}$-numéraire portfolio is therefore a driving risk-factor for the insurance premium.

In this section, we want to extend these results and consider more risk-factors which may influence the insurance premiums. To this end, we do not restrict to the case of a non time-varying intensity but want to explore the full power of stochastic intensity processes, given for a general $\mathbb{F}$-doubly stochastic Markov chain. In this setting, we do not obtain analytical expressions like in (2.22) or (2.23), but instead we apply Monte Carlo simulations to evaluate (2.6) for the special and most interesting case $t = 0$.

Given the setting of Section 2.1, we therefore assume X to follow a (general) $\mathbb{F}$-doubly stochastic Markov chain. In order to account for the investigation of covariate (or risk-factor) effects, we furthermore assume $\mathbb{F} = \mathbb{F}^Z$, with $\mathbf{Z} = ((Z^1(t), ..., Z^p(t))^\intercal)_{t\in[0,T]}$ a p-variate predictable process of covariates. Here, $\mathbf{Z}$ is assumed to represent, among others, individual-related as well as micro- and macro-economic risk factors, influencing the model as we will see more accurately in Section 2.3.1.

Due to Theorem A.7, for every $N \times N$-matrix-valued $\mathbb{F}$-adapted stochastic process $\widetilde{\mathbf{\Psi}}$, satisfying (A.4) and (A.5), there exists an $\mathbb{F}$-doubly stochastic Markov chain X, having intensity $\widetilde{\mathbf{\Psi}}$. Similarly to Assumption 2.2.1, we can therefore make the following assumption.

Assumption 2.3.1. *X is an $\mathbb{F}$-doubly stochastic Markov chain which admits the intensity* $\widetilde{\mathbf{\Psi}} = (\widetilde{\mathbf{\Psi}}_t)_{t\in[0,T]} = \left(\left[\psi_t^{j,k}\right]_{j,k\in\{1,2\}}\right)_{t\in[0,T]}$ *with*

$$-\psi_t^{j,j} = \psi_t^{j,k} = \psi^{jk}(t, Z^1(t), ..., Z^p(t)), \quad j, k \in \{1, 2\}, j \neq k, t \in [0, T], \quad (2.29)$$

for some measurable functions $\psi^{jk} : (\mathbb{R}^p, \mathcal{B}(\mathbb{R}^p)) \to (\mathbb{R}_+, \mathcal{B}(\mathbb{R}_+))$ with $\int_0^T \psi_s^{jk} ds < \infty$ for $j, k \in \{1, 2\}, j \neq k$.

Note that the proof of Theorem A.7 provides intrinsically the instruction, how to simulate paths of X given the paths of $\widetilde{\mathbf{\Psi}}$. An analogue but more detailed simulation scheme can be found in Bielecki and Rutkowski [24]. Here, the jump times $(\tau_j)_{j\geq 1}$, defined in (2.1) or (A.15), are obtained in terms of the intensities as

- $\tau_0 = 0$,

- $\tau_j = \inf\{\tau_{j-1} < t \leq T : e^{-\int_{\tau_{j-1}}^{t} \psi_s^{1+((j+1)\bmod 2),2-((j+1)\bmod 2)}ds} \leq U_j\}$, $j \geq 1$,   (2.30)

where $(U_j)_{j\geq 1}$ is a sequence of mutually independent uniformly distributed random variables on $[0, 1]$, which are also chosen to be independent of the intensity processes.

In our specific two-state setting with $X_0 = 1$ $\mathbb{P}$-a.s., any realization $(\tau_j(\omega))_{j\geq 1}$ determines the corresponding path $X(\omega)$ of the $\mathbb{F}$-doubly stochastic Markov chain. According to the path $X(\omega)$, the corresponding realization $D_T(\omega)$ of the cumulative payment process at time T defined in (2.5) can easily be derived. Given a corresponding realization $S^{\delta^*}(\omega)$ of the $\mathbb{P}$-numéraire portfolio, we can then apply Monte Carlo methods to obtain the insurance premium $\pi_0(D)$, given in (2.6) for $t = 0$. We will describe the simulation procedure in more detail in Section 2.3.6.

Because of Assumption (2.29), the covariate processes drive the randomness of the intensities and hence also determine the sequence $(\tau_j)_{j\in\mathbb{N}}$ of the jump times. In particular, if the $\mathbb{P}$-numéraire portfolio S^{δ^*} or its monthly returns, respectively, are considered as one of the covariate processes, we intrinsically model the required dependency structure between S^{δ^*} and the sequence of jump times. The other covariates can be any risk factors of interest such as individual related characteristics like sex, educational level, job sector, etc. or other external risk factors like monthly unemployment rates. In this way, we obtain a highly flexible premium determination framework for unemployment insurance products which can be adjusted according to various characteristics of the insured person and to external micro- and macro-economic risk-factors.

Note that this method can also be applied to capture the dependency structures affecting intensity processes in other kinds of applications.

In order to obtain reasonable insurance premiums, we have to estimate the intensity process $\widetilde{\Psi}$ accurately on a real dataset. In the following section we recall the setting of the well known Cox's proportional hazards model. Note that the estimators in this context are a priori estimators for the compensators of counting processes counting the jumps from employment to unemployment and from unemployment to employment in time. Hence, for the compensator-estimators to represent intensity-estimators of X, we first have to combine the two settings, an issue which, to the best of our knowledge, is missing (in the generality of stochastic compensators) in the literature.

2.3.1. Cox's Proportional Hazards Model and $\mathbb{F}$-Doubly Stochastic Markov Chains

The development of suitable statistical models for general multi-state switching processes with applications e.g. in biomedicine or econometrics, in which time-to-event variables are analyzed, has been an important task in statistics for a long time. Recently, there have been great efforts also on the implementational side, see Aalen et al. [1], Kneib and Hennerfeind [66], Simon et al. [88], Jackson [60] and de Wreede et al. [42].

The most frequently used model in survival analysis which is able to incorporate covariates belongs to the class of multiple hazard models and is known as the Cox model, see Cox [37, 38]. It is implemented in the R-function `coxph` from the R-package `survival`, see Therneau [91]. The function fits a Cox proportional hazards regression model including time dependent covariates, time dependent strata, and multiple events per subject.

The underlying estimators, implemented in these software packages, are based on the theory of multivariate counting processes and their compensators or, more precisely, on the compensators' densities with respect to the Lebesgue measure in the absolutely continuous case, see Andersen et al. [2]. This is due to the fact that any multi-state switching process generates a multivariate counting process by counting successively the transitions of the same type over time. An important question in this context is then, how the properties of the multi-state switching process and the counting process relate to each other. It is well known for a classical Markov chain that the compensator of the corresponding counting process is determined by the Markov chain's matrix-valued (deterministic) intensity function, see e.g. Andersen et al. [2]. Any estimator for the deterministic compensator is therefore implicitly an estimator for the intensity function of the Markov chain. The elaboration of suitable estimators for a counting process' compensator is based on a (partial) likelihood function which can be derived very generally for multivariate counting processes. Hence, it is natural to extend the statistical models to general multivariate counting processes with stochastic compensators of various forms, see Andersen et al. [2] or Therneau and Grambsch [92]. The remaining question is then whether it is still possible to characterize properties of the underlying multi-state switching processes like in the example of classical Markov chains.

Although estimators for stochastic compensators are used frequently and are often claimed to provide stochastic intensity processes for some underlying multi-state switching process, a thorough investigation on the relations between these processes is, to the best of our knowledge, missing in the statistical literature so far.

We now bridge this gap and show how stochastic compensators of the form presented by Cox's proportional hazards model, are naturally linked to a sub-class of F-doubly stochastic Markov chains with intensities by using the results of Theorem A.8. Therefore, the estimators which are implemented in the above mentioned software packages also represent suitable estimators for the matrix-valued stochastic intensity processes of the underlying F-doubly stochastic Markov chains.

We start by describing Cox's proportional hazards model as given in Andersen et al. [2, Section VII.2.] for the special case when successive jumps of n individuals are counted over time and any individual can perform two types of jump: from employment to unemployment and vice versa. The mathematical setting is as follows.

On the probability space $(\Omega, \mathcal{G}, \mathbb{P})$, we consider a $2n$-variate counting process

$$\mathbf{N} = ((\mathbf{N}(t))_{t \in [0,T]} = \left((N_1^{12}(t), N_1^{21}(t), ..., N_n^{12}(t), N_n^{21}(t))^\mathsf{T} \right)_{t \in [0,T]} \,,$$

where, $N_i^{jk}(t)$, $i \in \{1, ..., n\}$, $j, k \in \{1, 2\}$, $j \neq k$, is supposed to count the respective jumps of the i-th observed person from employment to unemployment denoted by the superscript "12", or from unemployment to employment denoted by the superscript "21", respectively, up to time t. The $2n$-variate process

$$\mathbf{\Lambda}(\boldsymbol{\theta}) = ((\mathbf{\Lambda}(t, \boldsymbol{\theta}))_{t \in [0,T]} = \left((\Lambda_1^{12}(t, \boldsymbol{\theta}), \Lambda_1^{21}(t, \boldsymbol{\theta}), ..., \Lambda_n^{12}(t, \boldsymbol{\theta}), \Lambda_n^{21}(t, \boldsymbol{\theta}))^{\mathsf{T}}\right)_{t \in [0,T]}$$

is assumed to be the compensator of $\mathbf{N}$ with respect to the filtration $\widehat{\mathbf{G}} = (\widehat{\mathcal{G}}_t)_{t \in [0,T]}$, with $\widehat{\mathcal{G}}_t = \mathcal{F}_t^N \vee \bigvee_{i=1}^n \mathcal{I}^i \vee \mathcal{A}$. Here, $\mathcal{F}_t^N = \sigma(\mathbf{N}(u) : u \leq t)$ is the σ-algebra generated by $\mathbf{N}$ up to time t, $\mathcal{I}^i$ is a σ-algebra related to individual i, and $\mathcal{A}$ some arbitrary σ-algebra. The filtration $\widehat{\mathbf{G}}$, interpreted as the level of information at any time $t \in [0, T]$, is hence determined by the information $\mathcal{F}_t^N$ generated by the observations of $\mathbf{N}$ up to and including time t, by additional i-specific information $\mathcal{I}^i$ of all individuals $i \in \{1, ..., n\}$ which is not changing over time, and by some other information $\mathcal{A}$ neither changing from individual to individual nor over time[3]. Note that the compensator is also supposed to depend on a parameter vector $\boldsymbol{\theta}$.

A desirable feature for statistical models is then that for every person $i \in \{1, ..., n\}$, the i-th component of $\mathbf{\Lambda}(\boldsymbol{\theta})$, i.e. the 2-variate process $\mathbf{\Lambda}_i(\boldsymbol{\theta}) = ((\Lambda_i^{12}(t, \boldsymbol{\theta}), \Lambda_i^{21}(t, \boldsymbol{\theta}))^{\mathsf{T}})_{t \in [0,T]}$, is given by the compensator of the counting processes $\mathbf{N}_i = ((N_i^{12}(t), N_i^{21}(t))^{\mathsf{T}})_{t \in [0,T]}$ with respect to the reduced filtration $\widehat{\mathbf{G}}^i = (\widehat{\mathcal{G}}_t^i)_{t \in [0,T]}$ with $\widehat{\mathcal{G}}_t^i = \mathcal{F}_t^{N_i} \vee \mathcal{I}^i \vee \mathcal{A}$. A sufficient condition for this to hold is that for every $t \in [0, T]$ the family $\left(\mathcal{F}_t^{N_i} \vee \mathcal{I}^i\right)_{i=1,...,n}$ is conditionally independent given $\mathcal{A}$. This yields a particular robustness of the statistical model since in this case the estimation of the compensator is independent of the number of observed individuals.

For every $i \in \{1, ..., n\}$, $j, k \in \{1, 2\}$, $j \neq k$, $t \in [0, T]$, we now assume the component $\Lambda_i^{jk}(t, \boldsymbol{\theta})$ of $\mathbf{\Lambda}(\boldsymbol{\theta})$ to have a density, i.e. a non-negative $\widehat{\mathbf{G}}$-predictable process $(\lambda_i^{jk}(t, \boldsymbol{\theta}))_{t \in [0,T]}$ such that

$$\Lambda_i^{jk}(t, \boldsymbol{\theta}) = \int_0^t \lambda_i^{jk}(s, \boldsymbol{\theta}) ds , \quad t \in [0, T] .$$

We then assume the densities to be of Cox's multiplicative form

$$\lambda_i^{jk}(t, \boldsymbol{\theta}) = Y_i^{jk}(t) \alpha_0^{jk}(t, \phi) e^{(\beta^{jk})^{\mathsf{T}} \mathbf{Z}_i(t)} , \tag{2.31}$$

where for $j, k \in \{1, 2\}$, $j \neq k$, Y_i^{jk} indicates whether individual i is at risk for a jump jk just before time $t \in [0, T]$; α_0^{jk} is the baseline-hazard function, unique for all individuals $i \in \{1, ..., n\}$; $\beta^{jk} = \left(\beta_1^{jk}, ..., \beta_p^{jk}\right)^{\mathsf{T}}$ is a parameter vector; $\boldsymbol{\theta}^{\mathsf{T}} = (\phi, (\beta^{12})^{\mathsf{T}}, (\beta^{21})^{\mathsf{T}})$ is the vector collecting all unknown parameters where ϕ is considered to be some nuisance parameter

---

[3] The authors in Andersen et al. [2] usually consider σ-algebras $\mathcal{F}_t^N$ and $\mathcal{A} = \mathcal{F}_0$, where $\mathcal{F}_0$ is supposed to contain information known at time $t = 0$. For the descriptions in this section, however, this setting needed to be extended. This further generalization has no consequence on the estimation results.

and $\mathbf{Z}_i = (\mathbf{Z}_i(t))_{t \in [0,T]} = ((Z_i^1(t), ..., Z_i^p(t))^\mathsf{T})_{t \in [0,T]}$ is the p-dimensional, predictable process of i-specific covariate processes as introduced above. Note that by (2.31) the densities depend only on $\boldsymbol{\theta}^{jk} := (\phi, (\boldsymbol{\beta}^{jk})^\mathsf{T}), j, k \in \{1, 2\}, j \neq k$, and consequently we can write $\Lambda_i^{jk}(t, \boldsymbol{\theta}^{jk})$ instead of $\Lambda_i^{jk}(t, \boldsymbol{\theta})$.

In order to connect this setting with F-doubly stochastic Markov chains, we now take $\mathcal{A} = \mathcal{F}_T^Z$, where $\mathbf{Z}(t) = [Z_i^l(t)]_{i \in \{1,...,n\}, l \in \{1,...,p\}}$ is the $n \times p$-(design) matrix valued stochastic process of all i-specific covariates $\mathbf{Z}_i, i \in \{1, \ldots, n\}$.

Recall that for every individual $i \in \{1, ..., n\}$ the process N_i^{jk}, $j, k \in \{1, 2\}, j \neq k$, counts the consecutive jumps from unemployment to employment and vice versa. Implicitly, one therefore observes the associated employment-unemployment progress of the i-th person, which we define by the right-continuous $\{1, 2\}$-valued stochastic process $X_i = (X_i(t))_{t \in [0,T]}$. If we set $H_i^j(t) := \mathbb{1}_{\{X_i(t)=j\}}$, we have

$$N_i^{jk}(t) = \int_0^t H_i^j(u-)dH_i^k(u) , \quad j, k \in \{1, 2\}, j \neq k,$$

and $\mathcal{F}_t^{X_i} = \mathcal{F}_t^{\mathbf{N}_i} \vee \sigma(X_i(0))$ for all $t \in [0, T]$.

An immediate consequence of Theorem A.8, particularly the equivalence between statement (i) and statement (iii), is then that every X_i follows an $\mathbb{F}^Z$-doubly stochastic Markov chain with 2×2-matrix valued intensity process $\boldsymbol{\Psi}^i$ having the entries

$$\psi_i^{j,k}(t, \boldsymbol{\theta}) = \alpha_0^{jk}(t, \phi)e^{(\boldsymbol{\beta}^{jk})^\mathsf{T}\mathbf{Z}_i(t)} , \qquad (2.32)$$

if and only if the compensators' densities of the counting processes N_i^{jk} are given as in (2.31) with $Y_i^{jk}(t) = H_i^j(t-)$ for $j, k \in \{1, 2\}, j \neq k$.

Moreover, since $\mathcal{F}_t^{X_i} = \sigma(X_i(0)) \vee \mathcal{F}_t^{\mathbf{N}_i}$, we have that for every $t \in [0, T]$ the family $(\mathcal{F}_t^{X_i})_{i=1,...,n}$ is conditionally independent given $\mathcal{F}_T^Z$, if and only if the family $(\mathcal{F}_t^{\mathbf{N}_i} \vee \mathcal{I}^i)_{i=1,...,n}$ is conditionally independent given $\mathcal{F}_T^Z$, where $\mathcal{I}^i = \sigma(X_i(0))$.

Hence, the estimation schemes for the individual counting processes $\mathbf{N}_i$ and their compensators $\boldsymbol{\Lambda}_i(\boldsymbol{\theta})$, which we briefly describe in Section 2.3.2, also provide estimators for the intensity matrix $\widetilde{\boldsymbol{\Psi}}^i$ of the underlying 2-state switching processes X_i if and only if for every individual $i \in \{1, ..., n\}$, the underlying employment-unemployment progress X_i is an $\mathbb{F}^Z$-doubly stochastic Markov chain, admitting intensity $\widetilde{\boldsymbol{\Psi}}^i$ and the family $(X_i)_{i=1,...,n}$ is conditionally independent given $\mathcal{F}_T^Z$.

Note that for the sake of simplicity, we focused in (2.31) on Cox's multiplicative form. The elaborated connection, however, obviously holds for any multiplicative form as given in Andersen et al. [2].

2.3.2. Estimators in Cox's Proportional Hazards Model

In this subsection we overview briefly the estimation procedure for the regression parameters $\boldsymbol{\beta}^{12}$ and $\boldsymbol{\beta}^{21}$, as well as for the underlying cumulative hazard functions in Cox's proportional hazard model. Following Andersen et al. [2], we do not suppose any specific form for the underlying baseline hazard functions $\alpha_0^{jk}(t, \phi), j, k \in \{1, 2\}$, $j \neq k$, appearing in (2.31). In particular, we assume no dependence of α_0^{jk} on a parameter ϕ and hence we write $\alpha_0^{jk}(t)$ instead of $\alpha_0^{jk}(t, \phi)$.

According to Andersen et al. [2], for $j \neq k \in \{1, 2\}$ and for fixed $\boldsymbol{\beta}^{jk}$ maximum likelihood estimates of the integrated underlying baseline hazard functions

$$A_0^{jk}(t) = \int_0^t \alpha_0^{jk}(u) du \tag{2.33}$$

are obtained by the so-called *Nelson-Aalen* estimators

$$\widehat{A}_0^{jk}(t, \boldsymbol{\beta}^{jk}) = \int_0^t \frac{J^{jk}(u)}{S_0^{jk}(\boldsymbol{\beta}^{jk}, u)} dN_\bullet^{jk}(u),$$

where $S_0^{jk}(\boldsymbol{\beta}^{jk}, t) := \sum_{i=1}^n \exp\{(\boldsymbol{\beta}^{jk})^\mathsf{T} \mathbf{Z}_i(t)\} H_i^j(t-)$ and $J^{jk}(t) = \mathbb{1}_{\{Y_\bullet^{jk}(t) > 0\}}$ with $Y_\bullet^{jk}(t) = \sum_{i=1}^n Y_i^{jk}(t)$ and $N_\bullet^{jk}(t) = \sum_{i=1}^n N_i^{jk}(t)$. Note that $J^{jk}(t)$ indicates if at least one individual is in risk of a transition of type "jk" at time t. Otherwise the function $S_0^{jk}(\boldsymbol{\beta}^{jk}, t)$ would be zero. Then the log *Cox partial likelihood*

$$\log L(\boldsymbol{\beta}^{jk}) = \sum_{i=1}^n \int_0^T (\boldsymbol{\beta}^{jk})^\mathsf{T} \mathbf{Z}_i(t) dN_i^{jk}(t) - \int_0^T \log S_0^{jk}(\boldsymbol{\beta}^{jk}, t) dN_\bullet^{jk}(t)$$

is maximized with respect to $\boldsymbol{\beta}^{jk}$, yielding $\widehat{\boldsymbol{\beta}}^{jk}$. The *Breslow estimator* for (2.33) is finally given by $\widehat{A}_0^{jk}(t, \widehat{\boldsymbol{\beta}}^{jk})$.

2.3.3. Description of the Dataset

We apply the estimators $\widehat{\boldsymbol{\beta}}^{jk}$ and $\widehat{A}_0^{jk}(t, \widehat{\boldsymbol{\beta}}^{jk})$ of Cox's proportional hazards model, which are implemented in the R-software package `survival`, to a dataset on integrated employment biographies provided by the IAB. In order to run the software, we first have to prepare this dataset appropriately.

The dataset of integrated labor market biographies (Stichprobe der integrierten Arbeitsmarktbiographien - SIAB) is a 2% sample of the population of the integrated employment biographies (IEB) of the IAB. We obtained access to the regional SIAB (Version 1975-2008) which is a file for scientific use in factual anonymous form that covers the employment and unemployment histories of 1,515,463 individuals in a total of 34,862,777 lines of data. The IEB comprises all individuals who showed one of the following statuses at least once during the observation period:

- employment subject to social security (recorded from 1975 onwards),

- marginal part-time employment (recorded from 1999 onwards),

Variable	Description
id	Identification number of individual
$Tstart$	Starting point of the time interval (in days)
$Tstop$	Endpoint of the time interval (in days)
$work$	Dummy variable for employment status (1=unemployed, 0=employed)
$status$	Dummy variable indicating whether a transition occurred or not; in the latter case, the observation is censored (1=yes, 0=no)
$trans$	Binary variable indicating for which transition an observation is at risk (1=employed to unemployed, 2=unemployed to employed)
sex	Dummy variable for gender (1=female, 0=male)
$birth$	Year of birth
$natio$	Nationality (1=German, 0=not German)
$occup^4$	Occupational status and working hours (categorical, 8 levels)
eco^4	Economic activity (categorical, 13 levels)
$state$	Federal state, where the individual is working (categorical, 18 levels[5])
$educ$	Educational level summarizing school education, school-leaving qualification and vocational training (categorical, 9 levels)
$wage$	Daily wage (in Euro; if more than one source at a time the average is used)
com	Dummy variable for commuter status (1=commuter, 0=no commuter)
$msci$	Monthly returns of the MSCI World Index (in %)
$urate$	Annual/monthly (federal state-specific) unemployment rate (in %)

Table 2.2.: Description of the variables of the integrated German labor market data for 1975-2008.

- receipt of benefits in accordance with Social Code Book III (recorded from 1975 onwards) or Social Code Book II (recorded from 2005 onwards),

- registered with the Federal Employment Agency as a jobseeker (recorded from 2000 onwards),

- planned or actual participation in an employment or training measure (recorded from 2000 onwards).

Note that all statuses are depicted exactly to the day. In general, the data have the following structure: each row represents an observation of a certain individual for a certain time interval, revealing some information about the working status of the individual together with several additional covariates concerning its educational level, sex, age, etc. An overview and a short description of all variables contained in the data is presented in Table 2.2, for more details see Dorner et al. [47]. The information regarding the working status allows a unique classification (in accordance with the German labor law) for each observation, whether an individual is employed or unemployed during the corresponding time interval.

---

[4]The variables *occup* and *eco* are not known during periods of unemployment. Hence, both variables are not considered in the intensity process for the transition from unemployment to employment.

[5]The 18 levels include the 16 German federal states themselves. Individuals with *no information* and individuals working in more than one federal state at the same time, are summarized in category *rest*.

In order to be able to apply the R-routine `coxph` on the SIAB data, several preparations are required that we describe in the following remark.

Remark 2.3.2.   *1. Since the data come from several records of different sources and are merged in the SIAB, the time intervals of observations belonging to one individual often overlap. Hence, in a first step we break down overlapping time intervals into disjoint intervals.*

  2. Information coming from different contemporaneous records of one individual are updated according to the most actual or crucial information. For example, if at least one of several contemporaneous records (coming from different sources) classifies an individual as a commuter, the individual is classified as a commuter for the corresponding time interval.

  3. The metric variable wage *is averaged over contemporaneous records; categorical variables with similar levels, coming from contemporaneous records, such as e.g. educational level (educ), economic activity (eco) or occupational status and working hours (occup), are merged.*

  4. As some of the variables are categorical with many factor levels, we have to aggregate some of them. For example for the variables economic activity (eco) *and* occupational status and working hours (occup) *only factor levels with at least 1,000,000 members are considered; factor levels with fewer members are summarized in the category* rest

  5. After aggregation of factor levels, still 29,778,075 lines of data, with no overlapping time intervals within individuals, remain.

  6. In addition to individual-related factors, two macro-economic variables are considered, the monthly returns of the MSCI World Index [6] *and the* annual/monthly unemployment rate [7] *(separately for the German federal states) for the period from 1975 to 2008. Since in the fitting procedure for the estimation of (2.32) all covariates have to be constant during each time interval, it has to be checked for every observation, if one of the two external variables has changed. In this case, the time interval has to be split into smaller ones where the external covariates are constant. Although both external variables only change monthly, still, this increases the dataset to contain more than 200,000,000 lines. Hence, we restrained to a data sample with the maximum size that can be handled by the R-routine, ending up with 150,000 individuals.*

To illustrate the structure of the data, an exemplary extraction is presented in Table 2.3.

---

[6]Source: http://www.msci.com/ .
[7]Source: http://www.arbeitsagentur.de/ .

id	Tstart	Tstop	status	trans	sex	state	...	msci	urate
25	7763	7793	0	1	0	Sachsen	...	0.027	18.3
25	7793	7854	0	1	0	Sachsen	...	0.027	18.6
25	7854	7874	1	1	0	Sachsen	...	0.080	18.3
25	7874	7935	0	2	0	Sachsen	...	0.018	17.7
25	7967	7998	0	2	0	Sachsen	...	-0.086	17.8
25	8003	8044	1	2	0	Sachsen	...	-0.067	17.8
25	8044	8073	0	1	0	Sachsen	...	0.072	20.4
25	8073	8078	0	1	0	Sachsen	...	0.033	20.4
37	5829	5844	0	1	1	Bayern	...	0.019	5.7
37	5844	5875	0	1	1	Bayern	...	0.034	5.1
37	5875	5903	0	1	1	Bayern	...	0.091	4.7
⋮	⋮	⋮	⋮	⋮	⋮	⋮	⋮	⋮	⋮

Table 2.3.: Structure of the SIAB data, exemplarily for a fictitious extraction.

2.3.4. Estimation Results for the Intensity Processes

In the following, we fit two (conditionally) independent Cox models for both transitions, including all covariates from Table 2.2. As the influence of the three metric covariates *birth*, *wage* and *urate* may be non-linear, we model their effect via cubic polynomials. For the metric regressor *msci* some special treatment becomes necessary. As the splitting procedure for *msci* creates almost exclusively time intervals within which the variable *msci* is equal for all observations, singularities are found in the design matrix with the consequence that the coxph function reports an error message. One possible solution to this problem is to include an interaction effect between the variables *state* and *msci*. Note that we do not face this problem for the variable *urate* as the time series are available separately for the German federal states. In the framework of the coxph package, the fit of the models can then be obtained by the following R-command, exemplarily for transition "12":

```
R> cox.obj <- coxph(Surv(Tstart, Tstop, status) ~ as.factor(sex) + birth
        + I(birth^2) + I(birth^3) + as.factor(natio)
        + as.factor(occup) + as.factor(eco) + as.factor(educ)
        + wage + I(wage^2) + I(wage^3) + as.factor(com)
        + as.factor(state) + as.factor(state):msci
        + urate + I(urate^2) + I(urate^3),
        data = sample, method = "breslow")
```

In Table 2.4, we present the estimated parameters $\widehat{\beta}^{12}$ and $\widehat{\beta}^{21}$.

The estimated cubic effects of the metric covariates *birth*, *wage*, and *urate*, exemplarily for transition "12", are illustrated in Figure 2.5. Especially for the variable *birth*, the effect is clearly non-linear. We find that for increasing date of birth the hazard rate from (2.32) has a cubic form: first it slightly decreases for older individuals, then increases for younger individuals, before it decreases again for very young individuals. In addition, higher wages substantially reduce the hazard rate. Furthermore, as expected, an increasing unemploy-

variable	$\hat{\beta}^{12}$	$\hat{\beta}^{21}$
sex:woman	-0.290	0.083
natio:german	-0.345	-0.208
natio:non-german	-0.483	-0.291
occup:skilled worker	-0.509	-
occup:untrained employee	-0.547	-
occup:employee	-0.930	-
occup:part-time employee (insured)	-1.167	-
occup:trainee	-1.807	-
occup:part-time employee (not insured)	-1.889	-
occup:rest	-1.250	-
birth	0.073	0.564
I(birth^2)	0.000	-0.003
I(birth^3)	-0.014	0.047
eco:retail	-0.169	-
eco:transport and communications	-0.370	-
eco:business oriented services	-0.171	-
eco:household oriented services	-0.021	-
eco:educational, social and health facility	-0.138	-
eco:public administration, social insurance	-0.182	-
eco:basic material and good production	-0.146	-
eco:structural steel, light-metal and machine engineering	-0.504	-
eco:steel forming, vehicle- and equipment engineering	-0.396	-
eco:consumer goods industry	-0.172	-
eco:main construction trades	0.597	-
eco:rest	-0.011	-
educ:abitur	0.878	2.460
educ:abitur and job training	0.679	0.828
educ:job training	1.365	1.860
educ:senior technical college (Fachhochschule)	0.869	1.051
educ:college/university	0.617	0.827
educ:elementary or secondary modern school, secondary school leaving certificate or similar	0.778	2.679
educ:elementary or secondary modern school, secondary school leyving certificate or similar with job training	0.582	2.787
educ:no education	2.853	2.495
wage	-0.854	1.380
I(wage^2)	-0.064	-0.148
I(wage^3)	0.052	-0.083
com:commuter	0.333	0.144
com:no commuter	0.090	0.147
state:Baden-Wuerttemberg	-1.362	-0.168
state:Bayern	-0.994	0.023
state:Berlin	-1.451	-0.077
state:Brandenburg	-1.533	0.002
state:Bremen	-1.354	-0.074
state:Hamburg	-1.410	-0.153
state:Hessen	-1.259	-0.157
state:Mecklenburg-Vorpommern	-1.454	0.009
state:Niedersachsen	-1.108	-0.018
state:Nordrhein-Westfalen	-1.382	-0.207
state:Rheinland-Pfalz	-1.201	-0.064
state:Saarland	-1.459	-0.236
state:Sachsen	-1.541	0.066
state:Sachsen-Anhalt	-1.582	0.017
state:Schleswig-Holstein	-1.096	0.034
state:Thueringen	-1.536	0.132
state:rest	-1.531	-0.407
urate	0.057	-0.022
I(urate^2)	0.002	-0.001
I(urate^3)	-0.000	0.000
state:no info:msci	0.346	-1.317
state:Baden-Wuerttemberg:msci	-1.342	-0.489
state:Bayern:msci	0.214	0.416
state:Berlin:msci	1.102	-0.532
state:Brandenburg:msci	-0.090	0.116
state:Bremen:msci	-1.860	-0.785
state:Hamburg:msci	-0.650	-0.865
state:Hessen:msci	-0.699	-0.960
state:Mecklenburg-Vorpommern:msci	-0.829	0.113
state:Niedersachsen:msci	0.046	-0.365
state:Nordrhein-Westfalen:msci	-0.957	-1.019
state:Rheinland-Pfalz:msci	-0.235	-1.277
state:Saarland:msci	-1.138	-1.370
state:Sachsen:msci	-0.285	-0.535
state:Sachsen-Anhalt:msci	-0.041	-0.294
state:Schleswig-Holstein:msci	-0.682	-0.077
state:Thueringen:msci	NA	NA
state:rest	NA	NA

Table 2.4.: Estimated linear effects with Cox's proportional hazards model.

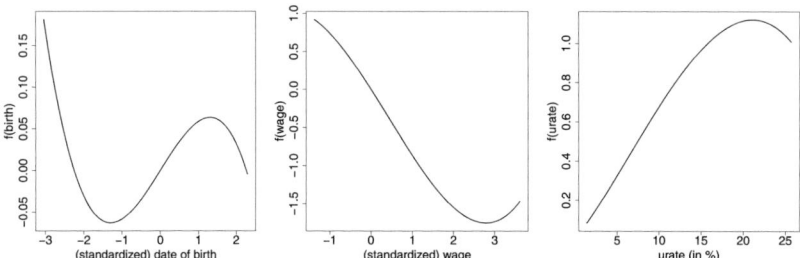

Figure 2.5.: Estimated non-linear effects of *birth, wage* and *urate* exemplarily for transition "12" (employment to unemployment).

ment rate in the federal state where an individual is employed increases its instantaneous probability of a transition into unemployment.

In the two upper plots of Figure 2.6 we illustrate the cumulative hazards for transition "12" (left) and for transition "21" (right) for an average individual based on the Nelson-Aalen estimates. Here, a "mean observation" represents an observation with constant covariate realizations obtained as the covariates averages over the used sample. We can see that as jumps from employment into unemployment occur much less often than vice versa and since individuals remain in employment usually much longer than in unemployment, the cumulative hazard for transition "21" (right) is much steeper than the one for transition "12" (left). The opposite effect is also observed for the survival functions, which are shown in the two lower plots of Figure 2.6.

2.3.5. Goodness-of-Fit Analysis

The estimation results and their graphical illustrations indicate that the model is reasonable and that the results coincide with intuitive anticipations. Nevertheless, statistically well-founded goodness-of-fit criteria are desirable. In this context, we first check the overall adequacy of the estimated model based on the Cox-Snell residuals, see Cox and Snell [39]. The Cox-Snell residual plots are shown in Figure 2.7 for both transitions. They show that the validity of the model is more disputable for transition "21" (unemployment to employment), since the estimated cumulative hazard of these residuals diverges from the bisecting line. One major reason might be that the two covariates *occup* and *eco* are not available for this transition and also the realizations of other variables, such as *state*, *wage* or *com*, are less reliable for individuals in the state "2" (unemployment), as they are predominantly extrapolated from the preceding employment period.

A test on the validity of the proportional hazards assumption versus the alternative of time-varying coefficients is provided by Grambsch and Therneau [56]. We show exemplarily the test results for the categorical covariate *state* and for transition "21" in Table 2.5. For

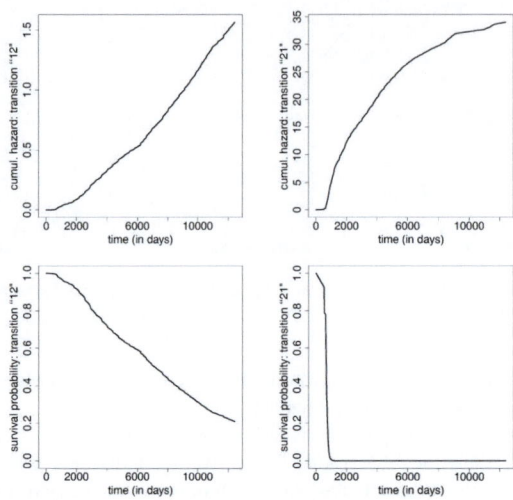

Figure 2.6.: Cumulative hazards (top) and the survival probabilities (bottom) for transition "12" (employment to unemployment, left) and for transition "21" (unemployment to employment, right) for
a "mean observation" based on the Nelson-Aalen estimates.

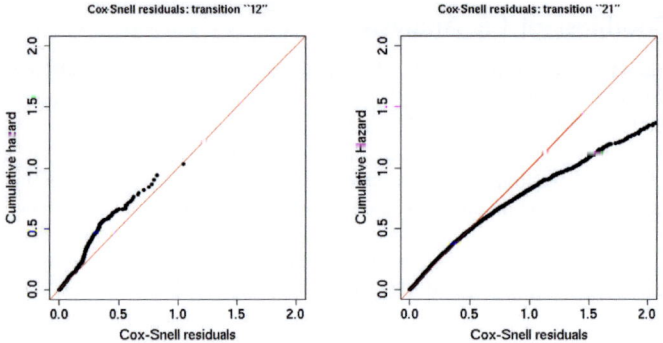

Figure 2.7.: Cox-Snell residuals for transition "12" (left) and "21" (right).

	rho	chisq	p
state:Baden-Wuerttemberg	-0.0047	2.4387	0.1184
state:Bayern	-0.0089	9.1932	0.0024
state:Berlin	0.0087	8.4570	0.0036
state:Brandenburg	0.0057	3.5889	0.0582
state:Bremen	0.0043	2.0088	0.1564
state:Hamburg	0.0083	7.7138	0.0055
state:Hessen	-0.0025	0.7137	0.3982
state:Mecklenburg-Vorpommern	0.0062	4.2591	0.0390
state:Niedersachsen	0.0053	3.3315	0.0680
state:Nordrhein-Westfalen	0.0048	2.8329	0.0924
state:Rheinland-Pfalz	0.0006	0.0424	0.8368
state:Saarland	0.0015	0.2595	0.6105
state:Sachsen	0.0059	3.9180	0.0478
state:Sachsen-Anhalt	0.0068	5.1429	0.0233
state:Schleswig-Holstein	0.0042	2.0377	0.1534
state:Thueringen	0.0053	3.1070	0.0780
state:sonst	0.0004	0.0134	0.9080

Table 2.5.: Test Results for Cox's Proportional Hazards Assumption

some categories, e.g. for *Bayern* and *Berlin*, the test results indicate significant non-proportionality. This can also be graphically illustrated by plots of the scaled Schoenfeld residuals, see Schoenfeld [82], versus a smoothed coefficient estimate, see Figure 2.8. Note that the left-continuous version of the Kaplan-Meier survival curve (without covariates) is used to scale the survival times. In general, we found violations of the proportional hazards assumption for several covariates and for both transitions. This indicates that several effects may vary over time and hence, models with predictors of a more complex structure could be worth of consideration. On the other hand, with regard to the large quantity of covariates it is not surprising that not all of them are consistent with the proportional hazards assumption. In the present framework of this section we ignore time-dependent effects, since our main objective of obtaining a good prediction for the underlying jump times is achieved in a very satisfactory way by the estimation results, as explained below.

Next, we investigate the robustness of the models. Due to the large size of our sample, the coefficient estimates are highly robust against slight changes in the data (critical limit is given by the range $[-0.1, 0.1]$). The approximate changes of the coefficient vector if single observations were dropped are illustrated in Figure 2.9, exemplarily for the coefficients of the variable *natio* and transition "12". In general, we found similar results for all other covariates and both transitions.

Finally we want to check the performance of our model in predicting transititon times by comparing true observed jump times with those estimated by the model. This is achieved by using a new goodness-of-fit method as follows.

Based on two different specifications for the estimated intensity processes, we can simulate realizations of a series of jump times for several individuals from our sample, by applying the simulation schemes for the jump times presented by (2.30). The first specifi-

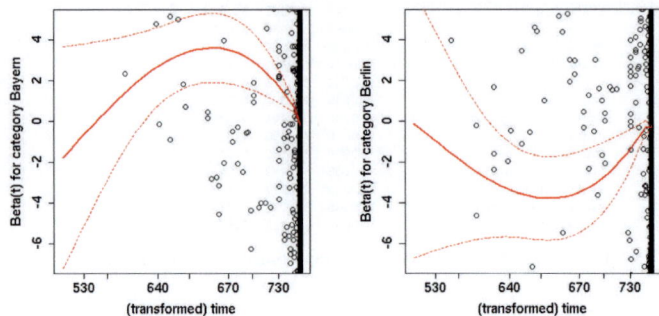

Figure 2.8.: Smoothed time-varying coefficients for the state categories *Bayern* (left) and *Berlin* (right) together with scaled Schoenfeld residuals, exemplarily for transition "21".

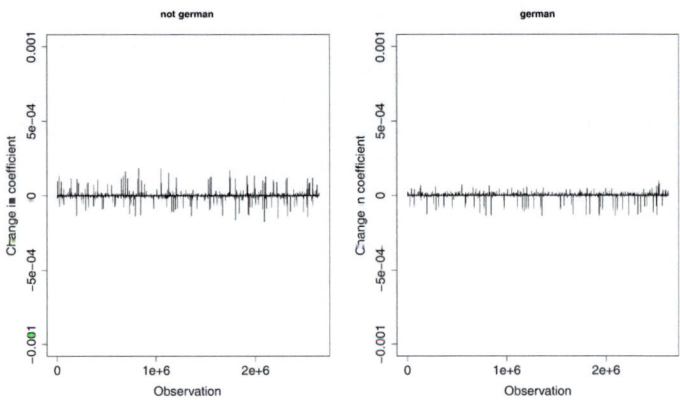

Figure 2.9.: Approximate changes of the coefficient vector if single observations were dropped, exemplarily for the coefficients of the variable *natio* and transition "12".

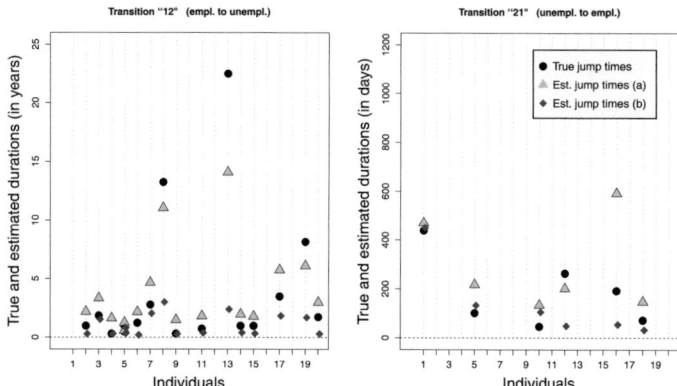

Figure 2.10.: Goodness-of-fit results: predicted durations in the states "1" (employment, left) and "2" (unemployment, right) for 20 individuals, both for Technique (a) and (b) and each based on 1,000 simulation runs, together with true durations.

cation uses the time-varying estimates of the baseline intensity functions as obtained from the R-routine. For individual i and $j, k \in \{1, 2\}, j \neq k$, we have

$$\hat{\psi}_i^{j,k}(t) = \hat{\alpha}_0^{jk}(t) \exp((\hat{\boldsymbol{\beta}}^{jk})^\mathsf{T} \mathbf{Z}_i(t)). \qquad \text{(Technique (a))}$$

However, the extrapolation of the time-varying estimated baseline hazards $\hat{\alpha}_0^{jk}$, as it would be necessary for predictions according to Technique (a), is generally a delicate task. Hence we consider a second specification based on constant baseline intensity estimates $\hat{\bar{\alpha}}_0^{jk}$, which are computed as the weighted mean of the time-varying baseline intensity estimates. For $j, k \in \{1, 2\}, j \neq k$, we obtain

$$\hat{\psi}_i^{j,k}(t) = \hat{\bar{\alpha}}_0^{jk} \exp((\hat{\boldsymbol{\beta}}^{jk})^\mathsf{T} \mathbf{Z}_i(t)). \qquad \text{(Technique (b))}$$

Technique (b) hence provides a more natural and simple way to predict jump times. Figure 2.10 shows the estimated and true durations in the states "1" (employment, left) and "2" (unemployment, right) for 20 individuals, both for Technique (a) and (b). It is seen that for both techniques the predicted jump times are reasonable and quite close to the true ones. We recognize that except for extraordinary long durations in the two states, Technique (b) with $\hat{\bar{\alpha}}_0^{12} = .00076$ and $\hat{\bar{\alpha}}_0^{21} = .00417$, performs quite well and is therefore used for the simulations in the next subsection.

Since the goodness-of-fit tests have shown the robustness of our analysis and the good predictive power of the model for jump times, we postpone further investigations of the proportional hazards assumption and possible extensions of the model to future research.

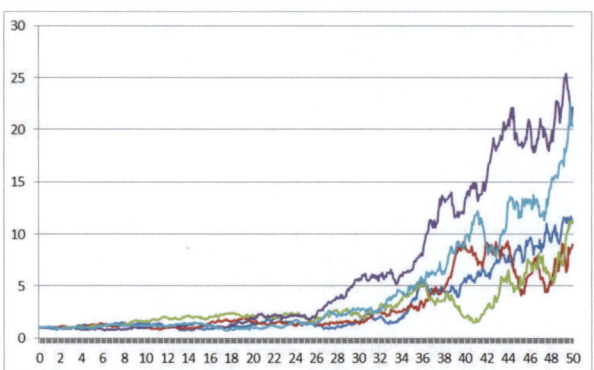

Figure 2.11.: Simulated paths of the $\mathbb{P}$-numéraire portfolio over a time horizon of $T = 50$ years.

2.3.6. Simulation Results for General $\mathbb{F}$-Doubly Stochastic Markov Chains

Taking advantage of the statistical analysis of Section 2.3.4, we are now able to compute the insurance premium for the unemployment insurance contracts as introduced in Section 2.1. More specifically, we want to approximate the unemployment insurance premium (2.6) by Monte Carlo simulations for $t = 0$. Therefore we need to simulate the underlying $\mathbb{F}$-doubly stochastic Markov chain with its corresponding jump times and the discounting factor, given by the $\mathbb{P}$-numéraire portfolio.

For the $\mathbb{P}$-numéraire portfolio we use a simulation scheme, described in detail in Ignatieva and Platen [59] and Platen and Rendek [79]. The underlying assumption is that the market follows the structure of a minimal market model as introduced in Platen [78]. In this setting, the $\mathbb{P}$-numéraire portfolio, discounted with the (locally) risk-less bank account, which for the sake of simplicity we describe by e^{rt} for some $r \geq 0$, is a squared Bessel process of dimension four and can be expressed as the sum of four independent time-transformed Wiener processes, see Platen and Heath [78]. Given the estimators for the model parameters in Ignatieva and Platen [59], we simulate $N = 10000$ paths of the $\mathbb{P}$-numéraire portfolio and choose $r = 2\%$. Five exemplary paths of the $\mathbb{P}$-numéraire portfolio are shown in Figure 2.11.

In order to obtain realizations of the $\mathbb{F}$-doubly stochastic Markov chain X, we first simulate the intensity processes $\widehat{\psi}^{1,2}$ and $\widehat{\psi}^{2,1}$ of the Markov chain's corresponding intensity matrix by assuming the concrete multiplicative structure $\widehat{\psi}^{1,2}(t) = \hat{\alpha}_0^{12} \exp\left((\widehat{\boldsymbol{\beta}}^{12})^\intercal \mathbf{Z}(t)\right)$ and $\widehat{\psi}^{2,1}(t) = \hat{\alpha}_0^{21} \exp\left((\widehat{\boldsymbol{\beta}}^{21})^\intercal \mathbf{Z}(t)\right)$, introduced as Technique (b) in Section 2.2.2, with $\hat{\alpha}_0^{12} = 0.0007611199$ and $\hat{\alpha}_0^{21} = 0.004170514$ the mean values of the estimated baseline hazards. Here, $\widehat{\beta}^{12}$ and $\widehat{\beta}^{21}$ are the estimated parameter vectors and $\mathbf{Z}$ is the vector of investigated covariates of Section 2.2.2.

For the sake of simplicity, we choose most of those covariates which may vary over time like *wage*, *urate*, or *state* as being constant over time although they could also be simulated according to some stochastic transition process. The only time-varying factor included in our simulations is the process of the monthly returns of our simulated paths of the $\mathbb{P}$-numéraire portfolio. In this setting we also assume for the simulations that the covariates $Z^1, ..., Z^p$ are independent. Note that this hypothesis is not required to perform the statistical analysis in Section 2.2.2.

Based on the resulting intensity processes we then simulate realizations of the series of jump times, underlying an $\mathbb{F}$-doubly stochastic Markov chain, as given in (2.30).

We then obtain realizations of an unemployment insurance contract's claim payments by testing if the insured person fulfills the criteria of receiving a claim payment at the respective payment dates. For the Monte Carlo approximations we perform this scheme of simulating the claim payments $N = 10,000$ times.

Similar to Section 2.2.2, we now present several plots in which we tested the simulated insurance premium' sensitivity against variation of its defining factors. To be more precise, the insurance premium (2.4) depends among others on

(i) contractual specifications, i.e. the maturity T, the deterministic installment plan c_i, $i \in \{1, ..., N\}$, the deferment period D, the waiting period W, and the requalification period R,

(ii) the categorial characteristics of the insured person, i.e. the covariate categories *sex*, *natio*, *occup*, *eco*, *state*, *educ*, and *commuter*,

(iii) the characteristics of the metric covariates *birth*, *wage*, *urate* and *msci*, i.e. the monthly returns of the simulated $\mathbb{P}$-numéraire portfolio for which the MSCI can be taken as a proxy.

In order to test the sensitivity of the premium, we fix the categorical and metric parameters at mean levels of the dataset and assume a contract where for $T = 10$ years a fictive payment $c_i = 1$ is paid to the policyholder in case of an insured event. We then let only the factor of interest vary.

Figure 2.12 shows the simulated insurance premiums for different maturities from $T = 3$ to $T = 43$ years and for different levels of the deferment, waiting, and requalification periods from 0 to 15 months, respectively. Both plots are coherent with the intuitive anticipations, for example that the insurance premium has to decline if the claim-excluding time periods increase. Note, however, that the effects of the deferment and waiting period are stronger than the effects of the requalification period.

Figure 2.13 shows the insurance premiums for the different levels of the covariates *sex* and *natio*. Based on our estimation results, the insurance premium for women and foreign employees are lower than for men or German employees.

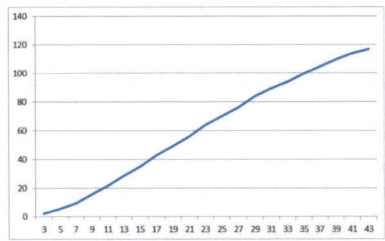 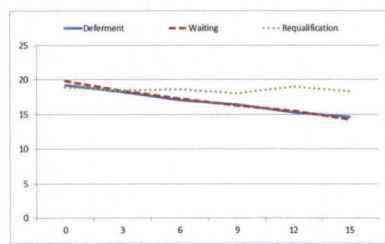

Figure 2.12.: Simulated insurance premiums for different maturities from $T = 3$ to $T = 43$ years (left) and for different levels of deferment, waiting and requalfication periods (right), verying from 0 to 15 months.

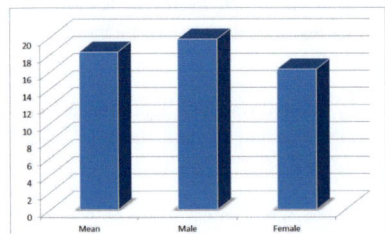 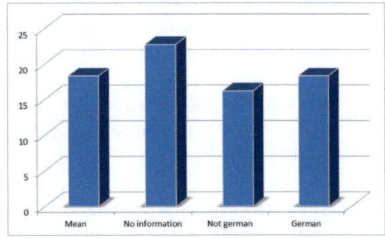

Figure 2.13.: Simulated insurance premium for the different categories of *sex* (left) and of the nationality *natio* (right).

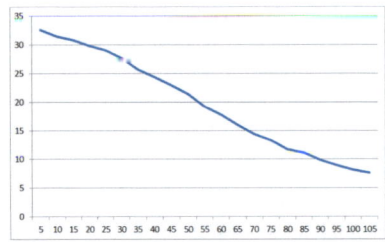 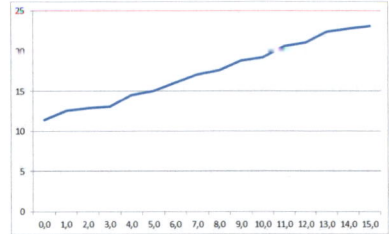

Figure 2.14.: Simulated insurance premiums for different levels of *wage* (left) of a (time-constant) *urate* (right).

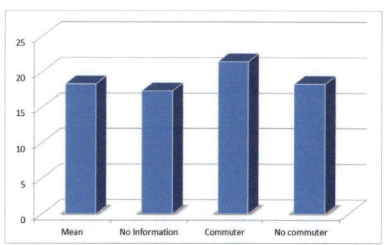 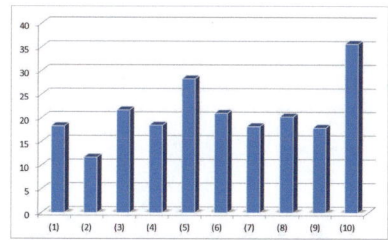

Figure 2.15.: Simulated insurance premiums for the different levels of *commuter* (left) and of *educ* (right), with (1) Sample mean, (2) No information, (3) Abitur, (4) Abitur and job training, (5) Job training, (6) Senior technical college (Fachhochschule), (7) College/University, (8) Elementary or secondary modern school, secondary school leaving certificate or similar, (9) Elementary or secondary modern school, secondary school leyving certificate or similar with job training, (10) No Education.

Figure 2.14 shows the simulated insurance premiums for different levels of the covariates *wage* and *urate*. It is seen that the higher is the wage, the lower would the insurance premium be. Moreover, the insurance premium obviously increases in times of a high unemployment rate.

Figure 2.15 shows the simulated insurance premiums for different levels of the covariates *commuter* and *educ*. Here we recognize that commuters generally have a higher insurance premium than non-commuters. Furthermore, people with no education would have to pay a significantly higher insurance premium than people with an education.

Figure 2.16 shows the simulated insurance premiums for different levels of the covariates *occup* and *eco*, and finally Figure 2.17 shows the simulated insurance premiums for different levels of the covariate *state*. Note that for the federal states, we have to adjust not only the covariate *state* but also *urate* and *msci*. For the unemployment rates, we take the contemporary valid unemployment rate and for *msci* the monthly returns of the simulated paths of the P-numéraire portfolio.

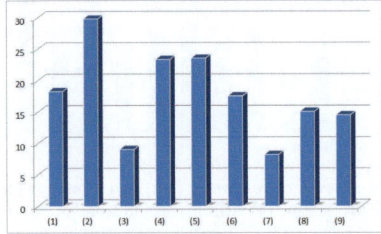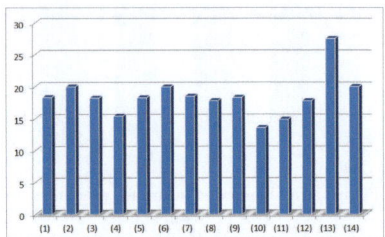

Figure 2.16.: Simulated insurance premiums for the different levels of *occup* (left) with (1) Sample mean,
(2) No information, (3) In education, (4) Un-/semi-skilled worker, (5) Skilled worker, (6)
Employee, (7) Part-time employee without unemployment insurance, (8) Part-time employee
with unemployment insurance, (9) Others, and of *eco* (right) with (1) Sample mean, (2) No
information, (3) Retail, (4) Transport and communications, (5) Business oriented services,
(6) Household oriented services, (7) Educational, social and health facility, (8) Public ad-
ministration, social insurance, (9) Basic material and good production, (10) Structural steel,
light-metal and machine engineering, (11) Steel forming, vehicle- and equipment engineering,
(12) Consumer goods industry, (13) Main construction trades, (14) Others.

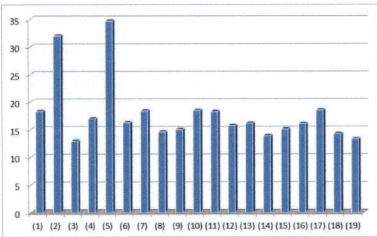

Figure 2.17.: Simulated insurance premiums for the different levels of *state* with (1) Sample mean, (2) No
information, (3) Baden-Württemberg, (4) Bayern, (5) Berlin, (6) Brandenburg, (7) Bre-
men, (8) Hamburg, (9) Hessen, (10) Mecklenburg-Vorpommern (11) Niedersachsen, (12)
Nordrhein-Westfalen, (13) Rheinland-Pfalz, (14) Saarland, (15) Sachsen, (16) Sachsen-
Anhalt, (17) Schleswig-Holstein, (18) Thüringen, (19) Others.

3. Mean-Variance Hedging for Mortality Claims with Longevity Bonds

3. Mean-Variance Hedging for Mortality Claims with Longevity Bonds

In this chapter, we treat a highly relevant topic of the life insurance and pension funds industry: the mitigation of longevity risk by trading in a longevity bond. We therefore study mean-variance hedging of a pure endowment, a term insurance and general life annuities by trading in a longevity bond with continuous rate payments proportional to the conditional survival probability. As stated in the introduction, we also discuss the introduction of a so called gratification annuity as an interesting insurance product for the life insurance market.

Following the framework of Section 1.1.2, the optimal hedging strategies are determined via their GKW-decompositions.

Our setting allows the further illustration of the results by assuming a general affine structure of the mortality intensity process and to compare paths of the hedging strategies as well as the residual hedging error for a gratification annuity and a simple life annuity with numerical simulations. This comparison additionally motivates the introduction of the gratification annuity on the life insurance market because it shows valuable features for the insurance industry.

Large parts of this chapter are based on the findings in Biagini, Rheinländer, and Widenmann [20].

3.1. The Reduced-Form Setting

We want to investigate mean-variance hedging for mortality claims in this chapter, and therefore consider the underlying process for an insured person of sojourning in the insurance policies' states to follow an $\mathbb{F}$-doubly stochastic Markov chain X with state-space $\{1, 2\}$. Here, the state 1 represents that the insured person is alive and the state 2 that he is dead. Then the state 2 must be absorbing according to Definition A.2 and as $t = 0$ in our context defines the beginning of the contract, we set $X_0 = 1$ $\mathbb{P}$-a.s.

In Appendix A.1 it is shown that these definitions cover reduced-form or hazard-rate models, provided they satisfy the immersion property, which are applied frequently in the context of credit risk or life insurance, see e.g. Barbarin [6] Bielecki and Rutkowski [24], Biagini and Schreiber [13] or Bielecki et al. [21, 22]. In order to be conform to the usual notations in the literature, we provide the results of this chapter directly within the reduced-form framework. The interested reader is then encouraged to reconsider the obtained results in the context of the definitions and results for $\mathbb{F}$-doubly stochastic Markov chains as given in Appendix A.

On $(\Omega, \mathcal{G}, \mathbb{P})$, we model the time of decease $\tau > 0$ of a person as a totally inaccessible random time with $P(\tau > t) > 0$ for any $t \in [0, T]$. Let $H = (H_t)_{t \in [0,T]}$ with $H_t := \mathbb{1}_{\{\tau \leq t\}}$ be the counting process of decease and $\mathbb{F}^H$ the filtration, generated by H. We suppose that the probability space supports also the augmented natural filtration $\mathbb{F}$ of some Brownian motion $W = (W_t)_{t \in [0,T]}$. In accordance with the $\mathbb{F}$-doubly stochastic Markov chain setting, we then let $\mathbb{F}$ be the reference filtration for our model, i.e. we assume $\mathbb{G} = \mathbb{F}^H \vee \mathbb{F}$. The following assumption is generally known as Hypothesis (H).

Assumption 3.1.1. (Hypothesis (H)) *Every $\mathbb{F}$-martingale remains a martingale in the larger filtration $\mathbb{G}$.*

Note that according to Assumption 3.1.1, W is a martingale in $\mathbb{G}$, and then by Lévy's characterization also a $(\mathbb{G}, \mathbb{P})$-Brownian motion.

Remark 3.1.2. *1) The statement of Assumption 3.1.1 is particularly equivalent to the following two characterizations:*

– *For every bounded, $\mathcal{F}_T$-measurable random variable Y, we have*

$$\mathbb{E}[Y \mid \mathcal{G}_t] = \mathbb{E}[Y \mid \mathcal{F}_t] \,,$$

– *for every $0 \le s \le t$, we have*

$$\mathbb{P}[\tau \le s \mid \mathcal{F}_t] = \mathbb{E}[\tau \le s \mid \mathcal{F}_T] \,,$$

see Brémaud and Yor [35] or Coculescu et al. [36].

2) Except the total inaccessibility of τ, all these assumptions follow implicitly if one assumes τ to be the first and only jump-time of an $\mathbb{F}$-doubly stochastic Markov chain, see Appendix A.1.

The conditional survival probability process $G = (G_t)_{t \in [0,T]}$, associated to τ, is defined as

$$G_t := P\left(\tau > t \mid \mathcal{F}_t\right) \,.$$

If we assume $G_t > 0$ for all $t \in [0,T]$, then we can define the $\mathbb{F}$-adapted *hazard process* $\Gamma = (\Gamma_t)_{t \in [0,T]}$ of the random time τ by

$$\Gamma_t := -\ln G_t \,. \tag{3.1}$$

Because τ is assumed to be totally inaccessible (and therefore avoids any $\mathbb{F}$-stopping time), the process Γ is continuous and due to Hypothesis (H) it is also increasing, see e.g. Coculescu et al. [36].

The counting process martingale $M = (M_t)_{t \in [0,T]}$, associated with H, is given as

$$M_t = H_t - \int_0^t (1 - H_u) \, d\Gamma_u \,,$$

see also (A.20). Recall that for every local $\mathbb{F}$-martingale N we have $[M, N] = 0$, see Proposition A.12 or Bielecki and Rutkowski [24]. In particular, we have $[M, W] = 0$.

Furthermore recall that the process $L = (L_t)_{t \in [0,T]}$ defined by $L_t := e^{\Gamma_t}(1 - H_t)$ is a martingale and satisfies $dL_t = -L_t dM_t$, see also (A.19) and (A.21).

As the focus of this chapter is to derive hedging strategies for mortality claims with respect to a particular hedging instrument, namely a longevity bond, we do not further comment on the underlying market specifications. For the sake of simplicity, we rather make the following simplifying assumption.

Assumption 3.1.3. *The measure $\mathbb{P}$ is a (local) martingale measure for the market, discounted with the discounting process $S^* = S^0$, where $S_t^0 = e^{rt}$, $t \in [0, T]$, for a fixed constant short rate $r \geq 0$.*

More generally, one could assume that the short rate is a stochastic process, independent of the mortality. For an adept study of this, see Dahl et al. [41].

Definition 3.1.4.   *1) A* pure endowment *is an insurance product which pays one unit of cash to the policyholder, given that the insured person is still alive at T, i.e.*

$$C^{pe} = \mathbb{1}_{\{\tau > T\}} = (1 - H_T).$$

2) A term insurance *is an insurance product which pays one unit to the policyholder (or his heirs) at the event of the insured person's decease, given that this happens before maturity. The cumulative payments up to time T are then given as*

$$C^{ti} = \mathbb{1}_{\{\tau \leq T\}} = H_T,$$

3) A general life annuity *is an insurance product where the policyholder receives general annuity payments as long as he is alive at a rate, given by some positive, bounded, $\mathbb{F}$-adapted stochastic process Y. The cumulative payments up to time T are then given as*

$$C^Y = \int_0^T (1 - H_u)\, Y_u \, du \,,$$

4) A gratification annuity *is a general life annuity with $Y_t = 1 - G_t$, $t \in [0, T]$, i.e.*

$$C^{ga} = \int_0^T (1 - H_u)(1 - G_u)\, du \,,$$

5) A simple life annuity *is a general life annuity with $Y_t = 1$, $t \in [0, T]$, i.e.*

$$C^{sl} = \int_0^T (1 - H_u) \, du \,,$$

Remark 3.1.5.   *1) A pure endowment C^{pe} is obviously a square-integrable T-contingent claim, according to Definition 1.1.4.*

2) *In order to account for the fact that a term insurance C^{ti} pays one unit at the time of decease τ, C^{ti} could more specifically be defined as a cumulative payment process $D^{ti} = (D_t^{ti})_{t \in [0,T]}$ with*

$$D_t^{ti} = H_t.$$

3) *Similarly, a general annuity C^Y is considered as the final value D_T^Y of a cumulative payment process $D^Y = (D_t^Y)_{t \in [0,T]}$ with*

$$D_t^Y = \int_0^t (1 - H_u) Y_u \, du. \tag{3.2}$$

With the numéraire $S_t^0 = e^{rt}$, Remark 3.1.5 and Definition 1.1.11, the discounted payments at or up to time T are given as.

$$\widehat{C}^{pe} := e^{-rT} \mathbb{1}_{\{\tau > T\}} = e^{-rT}(1 - H_T), \tag{3.3}$$

$$\widehat{C}^{ti} := \int_0^T e^{-ru} dH_u = e^{-r\tau} H_T, \tag{3.4}$$

$$\widehat{C}^Y := \int_0^T e^{-ru}(1 - H_u) Y_u \, du, \tag{3.5}$$

$$\widehat{C}^{ga} := \int_0^T e^{-ru}(1 - H_u)(1 - G_u) \, du, \tag{3.6}$$

$$\widehat{C}^{sl} := \int_0^T e^{-ru}(1 - H_u) \, du. \tag{3.7}$$

Note that the gratification annuity is a life insurance product which has, to the best of our knowledge, not yet been introduced on the life insurance market. As mentioned in the introduction, the conditional survival probability process G is usually inferred from a publicly accessible longevity index which describes the realized mortality of some representative collective showing similar characteristics to the insured person. As the continuous rate payments of a gratification annuity are given by $(1 - G_t) = \mathbb{P}(\tau \leq t \mid \mathcal{F}_t)$, a gratification annuity therefore rewards an insured person's higher longevity (e.g. due to healthier life style) as was originally expected.

Having sold a pure endowment, a term insurance, or a general annuity at time $t = 0$, an insurance company is exposed to mortality risk. In order to reduce this risk exposure, the insurance company is interested in purchasing assets showing similar characteristics to its contracts. Therefore, we assume that it is possible to trade on the financial market in an instrument called a longevity bond.

Definition 3.1.6. *A longevity bond is a contract which pays a declining rate, given by the conditional survival probability process G. It can be represented by the cumulative payment process $B = (B_t)_{t \in [0,T]}$ with*

$$B_t = \int_0^t G_u \, du.$$

The discounted longevity bond is then represented by the discounted cumulative payment process $\widehat{B} = (\widehat{B}_t)_{t \in [0,T]}$ with

$$\widehat{B}_t = \int_0^t e^{-ru} G_u \, du. \tag{3.8}$$

Note that the payment, generated by this (coupon based) bond, has also the form of an annuity where the declining rate is given by the conditional survival probability for the age cohort of the insured person. It does not depend on his individual life history, in contrast to the payouts of the considered mortality claims.

In the context of Definition 1.1.19 and its interpretation in Remark 1.1.20, the discounted value process $\widehat{U}^{lb} = (\widehat{U}_t^{lb})_{t \in [0,T]}$, associated with the longevity bond is given by

$$\widehat{U}_t^{lb} = \mathbb{E}\left[\int_0^T e^{-ru} G_u \, du \,\middle|\, \mathcal{G}_t \right]. \tag{3.9}$$

Note again that we implicitly assume $\mathbb{P}$ to be some pricing measure reflecting the market price of risk.

Our goal is now to hedge the risk exposure from having sold either a pure endowment, a term insurance or a general annuity by trading dynamically in the longevity bond with discounted intrinsic value process $\widehat{U}^{lb}$.

3.2. Mean-Variance Hedging for a Single Life Status

We start the analysis of optimal hedging strategies by considering a single life status. To this end, we make the following assumption.

Assumption 3.2.1. The hazard process Γ provides

$$e^{\Gamma_T} \in L^2(\mathbb{P}). \tag{3.10}$$

Definition 3.2.2. A simple claim is a random variable of the form $(1 - H_T) Z$ for some integrable $\mathcal{F}_T$-measurable random variable Z. By martingale representation with respect to the Brownian filtration $\mathbb{F}^W$, there exists a process $\psi \in L(W)$ such that

$$\mathbb{E}\left[e^{-\Gamma_T} Z \mid \mathcal{F}_t \right] = c^Z + \int_0^t \psi_s \, dW_s, \tag{3.11}$$

with $c^Z = \mathbb{E}\left[e^{-\Gamma_T} Z \right]$.

For simple claims, a martingale representation can then be found by integration by parts as in the proof in Bielecki and Rutkowski [24, Proposition 5.2.2].

Proposition 3.2.3. *Let $X = (1 - H_T) Z$, $T \geq 0$, be a simple claim. Then*

$$\mathbb{E}\left[X \mid \mathcal{G}_t \right] = c^X + \int_0^t \zeta_s^W \, dW_s + \int_{0+}^t \zeta_s^M \, dM_s \tag{3.12}$$

where $c^X = \mathbb{E}[X] = \mathbb{E}\left[e^{-\Gamma_T} Z \right]$, $\zeta_s^W = L_{s-} \psi_s$ and $\zeta_s^M = -L_{s-} \left(c^Z + \int_0^t \psi_s \, dW_s \right)$.

Obviously, the discounted pure endowment $\widehat{C}^{pe}$ is a simple claim. However, both discounted term insurance as well as discounted general annuities have to be dealt with differently.

Remark 3.2.4. *A general representation result, see Bielecki and Rutkowski [24, Proposition 5.2.2], can be obtained by approximating $\mathcal{G}_T$-measurable random variables in $L^2(P)$ by simple claims and by using the fact that the spaces of $L^2(W)$ and $L^2(M)$ are closed in $L^2(P)$. The result then states that each square-integrable $(\mathcal{G}_t)$-martingale N can be written as*

$$N_t = N_0 + \int_0^{\tau \wedge t} \zeta_s^W \, dW_s + \int_{0+}^{\tau \wedge t} \zeta_s^M \, dM_s \, ,$$

for integrands $\zeta^W \in L^2(W)$ and $\zeta^M \in L^2(M)$. It does not, however, give any information how to calculate the integrands for claims which are not simple.

Proposition 3.2.5. *The GKW-decomposition for the discounted intrinsic value process $\widehat{U}^{pe} = (\widehat{U}_t^{pe})_{t \in [0,T]}$ of a pure endowment with respect to W is*

$$\widehat{U}_t^{pe} = \mathbb{E}\left[e^{-rT} (1 - H_T) \mid \mathcal{G}_t \right] = c^{pe} + \int_0^t \alpha_s^W \, dW_s + \int_{0+}^t \alpha_s^M \, dM_s \, , \tag{3.13}$$

where the predictable integrands α^W and α^M are given as

$$\alpha_s^W = L_{s-} \psi_s \, , \tag{3.14}$$

$$\alpha_s^M = -L_{s-} \left(c^{pe} + \int_0^s \psi_v \, dW_v \right) \, ,$$

with $c^{pe} = \mathbb{E}\left[e^{-rT} e^{-\Gamma_T} \right]$. Here ψ corresponds to the integrand in (3.11) for the choice $Z = e^{-rT}$.

Proof. $\widehat{C}^{pe}$ is a square-integrable simple claim with $Z = e^{-rT}$, hence Decomposition (3.12) holds. Because α^W and α^M are $\mathbb{G}$-predictable, because the Brownian motion W is a square integrable martingale and because the (local) martingale $\left(\int_{0+}^t \alpha_s^M dM_s \right)_{t \in [0,T]}$ is strongly orthogonal to W, Lemma B.3 yields the result. $\square$

We turn now our attention to term insurance. For completeness, we provide its GKW-decomposition in our setting. The results could also be derived by applying results of Barbarin [5].

Let us first observe that by martingale representation there exists a process $\chi \in L^2(W)$ such that

$$\mathbb{E}\left[\int_0^T e^{-ru}e^{-\Gamma_u}\,d\Gamma_u \,\middle|\, \mathcal{F}_t\right] = c^{ti} + \int_0^t \chi_u\,dW_u, \tag{3.15}$$

with $c^{ti} = \mathbb{E}\left[\int_0^T e^{-ru}e^{-\Gamma_u}\,d\Gamma_u\right]$.

Proposition 3.2.6. *The GKW-decomposition for the discounted intrinsic value process $\widehat{U}^{ti} = (\widehat{U}^{ti}_t)_{t\in[0,T]}$ of a term insurance with respect to W is*

$$\widehat{U}^{ti}_t = \mathbb{E}\left[e^{-r\tau}H_T \mid \mathcal{G}_t\right] = c^{ti} + \int_0^t \beta^W_s\,dW_s + \int_{0+}^t \beta^M_s\,dM_s, \tag{3.16}$$

where the predictable integrands β^W and β^M are given as

$$\beta^W_s = L_{s-}\chi_s, \tag{3.17}$$
$$\beta^M_s = e^{-r(s\wedge\tau)} - L_{s-}\left(c^{ti} + \int_0^s \chi_v\,dW_v - \int_0^s e^{-rv}e^{-\Gamma_v}\,d\Gamma_v\right),$$

with $c^{ti} = \mathbb{E}\left[\int_0^T e^{-ru}e^{-\Gamma_u}\,d\Gamma_u\right]$.

Proof. We write

$$\mathbb{E}\left[e^{-r\tau}H_T \mid \mathcal{G}_t\right] = H_t\mathbb{E}\left[e^{-r\tau}H_T \mid \mathcal{G}_t\right] + (1-H_t)\mathbb{E}\left[e^{-r\tau}H_T \mid \mathcal{G}_t\right] \tag{3.18}$$

and find the canonical decompositions of the two terms on the right hand side into a local martingale and a finite variation part separately.

Since $H_tH_T = H_t$, and $H_te^{-r\tau}$ is $\mathcal{G}_t$-measurable, we get for the first term by integration by parts

$$\begin{aligned}
H_t\mathbb{E}\left[e^{-r\tau}H_T \mid \mathcal{G}_t\right] = H_te^{-r\tau} &= H_te^{-r(t\wedge\tau)} \\
&= 0 + \int_0^t H_{s-}\,de^{-r(s\wedge\tau)} + \int_0^t e^{-r(s\wedge\tau)}\,dH_s \\
&= 0 + \int_0^t e^{-r(s\wedge\tau)}\,dM_s + X^1_t, \tag{3.19}
\end{aligned}$$

where

$$X^1_t = \int_0^t e^{-r(s\wedge\tau)}(1-H_s)\,d\Gamma_s = \int_0^t e^{-rs}(1-H_s)\,d\Gamma_s\,.$$

Note that we used here that $H_s = 0$ on $[0, \tau)$, and that $H_s = M_s + \Gamma_{s \wedge \tau}$, see (A.20).
For the second term of the r.h.s in (3.18), we get by Corollary A.15 that

$$(1 - H_t) \, \mathbb{E} \left[e^{-r\tau} H_T \mid \mathcal{G}_t \right] = (1 - H_t) \, \mathbb{E} \left[\int_t^T e^{-rs} e^{\Gamma_t - \Gamma_s} \, d\Gamma_s \, \Big| \, \mathcal{F}_t \right]$$

$$= L_t \mathbb{E} \left[\int_t^T e^{-rs} e^{-\Gamma_s} \, d\Gamma_s \, \Big| \, \mathcal{F}_t \right].$$

Again by integration by parts as well as the martingale representation (3.15),

$$L_t \mathbb{E} \left[\int_t^T e^{-rs} e^{-\Gamma_s} \, d\Gamma_s \, \Big| \, \mathcal{F}_t \right] = L_t \left(\mathbb{E} \left[\int_0^T e^{-rs} e^{-\Gamma_s} \, d\Gamma_s \, \Big| \, \mathcal{F}_t \right] - \int_0^t e^{-rs} e^{-\Gamma_s} \, d\Gamma_s \right)$$

$$= c^{ti} + \int_0^t \phi_s \, dW_s + \int_{0+}^t \nu_s \, dM_s + X_t^2, \qquad (3.20)$$

where

$$\phi_s = L_{s-} \chi_s,$$

$$\nu_s = -L_{s-} \left(c^{ti} + \int_0^s \chi_v \, dW_v - \int_0^s e^{-rv} e^{-\Gamma_v} \, d\Gamma_v \right),$$

and

$$X_t^2 = - \int_0^t L_s e^{-rs} e^{-\Gamma_s} d\Gamma_s = -X_t^1.$$

The result now follows by combining (3.19) and (3.20) and with Lemma B.3, since β^W and β^M are $\mathbb{G}$-predictable and W is a square integrable martingale. $\qquad \square$

Now we turn to a general annuity. As stated in the introduction, the following decompositions could also be derived by applying results in Barbarin [5] or Biagini and Cretarola [12] to our setting. Here they are computed more explicitly under our model assumptions.

By martingale representation, for each $u \in [0, T]$ there exists a predictable process $(\theta_{u,s}^Y)_{s \in [0,T]} \in L^2(W)$, with $\theta_{u,s}^Y = 0$ if $s > u$, such that

$$\mathbb{E} \left[e^{-ru} Y_u e^{-\Gamma_u} \mid \mathcal{F}_t \right] = c_u^Y + \int_0^{t \wedge u} \theta_{u,s}^Y \, dW_s$$

$$= c_u^Y + \int_0^t \theta_{u,s}^Y \mathbb{1}_{[0,u]}(s) \, dW_s, \qquad (3.21)$$

where $c_u^Y = \mathbb{E} \left[e^{-ru} Y_u e^{-\Gamma_u} \right]$. We set $c^Y = \int_0^T c_u^Y \, du < \infty$. Note that the processes $\left(\int_0^t \theta_{u,s} \mathbb{1}_{[0,u]}(s) dW_s \right)_{t \in [0,T]}$ are bounded martingales, uniformly in u.

Proposition 3.2.7. *The GKW-decomposition of the discounted intrinsic value process* $\widehat{U}^Y = (\widehat{U}^Y_t)_{t \in [0,T]}$ *of a general annuity with respect to W is*

$$\widehat{U}^Y_t = \mathbb{E}\left[\int_0^T e^{-ru}\left(1 - H_u\right)Y_u\,du \,\Big|\, \mathcal{G}_t\right] = c^Y + \int_0^t \rho^W_s\,dW_s + \int_{0+}^t \rho^M_s\,dM_s, \qquad (3.22)$$

where the predictable integrands ρ^W and ρ^M are given as

$$\rho^W_s = L_{s-}\int_s^T \theta^Y_{u,s}\,du, \qquad (3.23)$$

$$\rho^M_s = -L_{s-}\int_s^T \left(c^Y_u + \int_0^s \theta^Y_{u,v}\,dW_v\right)du .$$

Proof. By Corollary A.22, we have for $u \in (t, T]$

$$\mathbb{E}\left[e^{-ru}\left(1 - H_u\right)Y_u \,\big|\, \mathcal{G}_t\right] = (1 - H_t)\,\mathbb{E}\left[e^{-ru}Y_u e^{\Gamma_t - \Gamma_u} \,\big|\, \mathcal{F}_t\right]$$
$$= L_t\mathbb{E}\left[e^{-ru}Y_u e^{-\Gamma_u} \,\big|\, \mathcal{F}_t\right].$$

For $u \in [0, t]$ we have

$$\mathbb{E}\left[e^{-ru}\left(1 - H_u\right)Y_u \,\big|\, \mathcal{G}_t\right] = (1 - H_u)\,\mathbb{E}\left[e^{-ru}Y_u \,\big|\, \mathcal{F}_t\right]$$
$$= L_u\mathbb{E}\left[e^{-ru}Y_u e^{-\Gamma_u} \,\big|\, \mathcal{F}_t\right].$$

Hence, for every $u \in [0, T]$ we have

$$\mathbb{E}\left[e^{-ru}\left(1 - H_u\right)Y_u \,\big|\, \mathcal{G}_t\right] = L_{u \wedge t}\mathbb{E}\left[e^{-ru}Y_u e^{-\Gamma_u} \,\big|\, \mathcal{F}_t\right]$$
$$= L^u_t\mathbb{E}\left[e^{-ru}Y_u e^{-\Gamma_u} \,\big|\, \mathcal{F}_t\right],$$

where L^u is the process L stopped at time $u \in [0, T]$. By integration by parts and (3.21),

$$L^u_t\mathbb{E}\left[e^{-ru}Y_u e^{-\Gamma_u} \,\big|\, \mathcal{F}_t\right] = c^Y_u + \int_0^t \phi_{u,s}\,dW_s + \int_{0+}^t \nu_{u,v}\,dM_s, \qquad (3.24)$$

where

$$\phi_{u,s} = L_{s-}\,\theta^Y_{u,s}\mathbb{1}_{[0,u]}(s),$$

$$\nu_{u,s} = -L_{s-}\left(c^Y_u + \int_0^s \theta^Y_{u,v}\mathbb{1}_{[0,u]}(v)\,dW_v\right)\mathbb{1}_{[0,u]}(s).$$

By Fubini, as well as the Itô-isometry,

$$\mathbb{E}\left[\int_0^T\int_0^T \phi^2_{u,s}\,du\,ds\right] = \int_0^T \mathbb{E}\left[\int_0^T \phi^2_{u,s}\,ds\right]du$$

$$= \int_0^T \mathbb{E}\left[\left(\int_0^T \phi_{u,s}\, dW_s \right)^2 \right] du$$

$$\leq C_1 T,$$

where, by Lemma 3.2.8 below, we have

$$C_1 = \sup_{0 \leq u \leq T} \left\| \int_0^T \phi_{u,s}\, dW_s \right\|_{L^2}^2 < \infty.$$

Moreover,

$$\mathbb{E}\left[\int_0^T \int_0^T \nu_{u,s}^2\, du\, d\Gamma_s \right] = \int_0^T \mathbb{E}\left[\int_0^T \nu_{u,s}^2\, d\Gamma_s \right] du$$

$$= \int_0^T \mathbb{E}\left[\left(\int_0^T \nu_{u,s}\, dM_s \right)^2 \right] du$$

$$\leq C_2 T,$$

where, by Lemma 3.2.9 below, we have

$$C_2 = \sup_{0 \leq u \leq T} \left\| \int_0^T \nu_{u,s}\, dM_s \right\|_{L^2}^2 < \infty.$$

Hence we may apply the stochastic Fubini theorem, as given e.g. in Protter [80, Theorem IV.65] to get from (3.24)

$$\mathbb{E}\left[\int_0^T e^{-ru}(1 - H_u) Y_u\, du \,\middle|\, \mathcal{G}_t \right] = \int_0^T \mathbb{E}\left[e^{-ru}(1 - H_u) Y_u \,\middle|\, \mathcal{G}_t \right] du$$

$$= \int_0^T \left(c_u^Y + \int_0^t \phi_{u,s}\, dW_s + \int_{0+}^t \nu_{u,s}\, dM_s \right) du$$

$$= c^Y + \int_0^t \int_0^T \phi_{u,s}\, du\, dW_s + \int_{0+}^t \int_0^T \nu_{u,s}\, du\, dM_s$$

$$= c^Y + \int_0^t \rho_s^W\, dW_s + \int_{0+}^t \rho_s^M\, dM_s,$$

where the predictable integrands ρ^W, ρ^M are as desired. The result now follows with Lemma B.3, since W is a square integrable martingale. $\qquad \square$

Lemma 3.2.8. *We have*

$$\sup_{0 \leq u \leq T} \left\| \int_0^T \phi_{u,s}\, dW_s \right\|_{L^2} < \infty. \tag{3.25}$$

Proof. First note that since $(1 - H_{s-}) \leq 1$, we have

$$\mathbb{E}\left[\int_0^T L_{s-}^2 \theta_{u,s}^2 \mathbb{1}_{[0,u]}(s)\, ds\right] \leq \mathbb{E}\left[\int_0^T e^{2\Gamma_s}\theta_{u,s}^2 \mathbb{1}_{[0,u]}(s)\, ds\right].$$

By the Itô-isometry, (3.25) therefore holds if

$$\mathbb{E}\left[\left(\int_0^T e^{\Gamma_s}\theta_{u,s}\mathbb{1}_{[0,u]}(s)\, dW_s\right)^2\right] < \infty. \tag{3.26}$$

Since by (3.21) the $\left(\int_0^t \theta_{u,s}\mathbb{1}_{[0,u]}(s)dW_s\right)_{t\in[0,T]}$ are bounded martingales, uniformly in u, we have by integration by parts that for each $u \in [0,T]$ that

$$\left|\int_0^T e^{\Gamma_s}\theta_{u,s}\mathbb{1}_{[0,u]}(s)\, dW_s\right| = \left|e^{\Gamma_T}\int_0^T \theta_{u,s}\mathbb{1}_{[0,u]}(s)\, dW_s - \int_0^T \int_0^s \theta_{u,v}\mathbb{1}_{[0,u]}(v)\, dW_v\, de^{\Gamma_s}\right|$$
$$\leq 2Ce^{\Gamma_T},$$

where the constant C is independent of u. Therefore (3.26) follows from assumption (3.10), namely that $e^{\Gamma_T} \in L^2(\mathbb{P})$. $\square$

Lemma 3.2.9. *We have*

$$\sup_{0\leq u\leq T}\left\|\int_0^T \nu_{u,s}\, dM_s\right\|_{L^2} < \infty. \tag{3.27}$$

Proof. As

$$\nu_{u,s} = -L_{s-}\left(c_u + \int_0^s \theta_{u,v}\mathbb{1}_{[0,u]}(v)\, dW_v\right)\mathbb{1}_{[0,u]}(s),$$

and $\left(\int_0^t \theta_{u,s}\mathbb{1}_{[0,u]}(s)dW_s\right)_{t\in[0,T]}$ are bounded martingales, uniformly in u, it follows that the $\nu_{u,\cdot}$ are bounded as well by some constant C independent of u. Therefore

$$\mathbb{E}\left[\int_0^T \nu_{u,s}^2\, d\langle M\rangle_s\right] \leq C^2\mathbb{E}\left[\langle M\rangle_T\right] = C^2\mathbb{E}\left[\Gamma_{T\wedge\tau}\right] \leq C^2\mathbb{E}\left[\Gamma_T\right],$$

such that the Itô isometry as well as (3.10) prove the assertion. $\square$

We have already introduced a new type of insurance product, namely a gratification annuity which we think of as an interesting insurance product for the life insurance market. In order to compare this product to an existing annuity, we also consider the simple life annuity. The GKW-decompositions are given in the following corollary.

Corollary 3.2.10. *The GKW-decompositions of the discounted intrinsic value processes* $\widehat{U}^{ga} = (\widehat{U}_t^{ga})_{t\in[0,T]}$ *and* $\widehat{U}^{la} = (\widehat{U}_t^{la})_{t\in[0,T]}$ *of a gratification annuity and a simple life annuity with respect to W are*

$$\widehat{U}_t^{ga} = \mathbb{E}\left[\int_0^T e^{-ru}\left(1 - H_u\right)\left(1 - G_u\right) du \,\Big|\, \mathcal{G}_t\right] = c^{ga} + \int_0^t \gamma_s^W \, dW_s + \int_{0+}^t \gamma_s^M \, dM_s\,, \quad (3.28)$$

$$\widehat{U}_t^{la} = \mathbb{E}\left[\int_0^T e^{-ru}\left(1 - H_u\right) du \,\Big|\, \mathcal{G}_t\right] = c^{la} + \int_0^t \delta_s^W \, dW_s + \int_{0+}^t \delta_s^M \, dM_s\,, \quad (3.29)$$

where the predictable integrands γ^W and γ^M as well as δ^W and δ^M are given as

$$\gamma_s^W = L_{s-} \int_s^T \theta_{u,s}^{ga} \, du\,, \quad (3.30)$$

$$\gamma_s^M = -L_{s-} \int_s^T \left(c_u^{ga} + \int_0^s \theta_{u,v}^{ga} \, dW_v\right) du\,, \quad (3.31)$$

and

$$\delta_s^W = L_{s-} \int_s^T \theta_{u,s}^{la} \, du\,, \quad (3.32)$$

$$\delta_s^M = -L_{s-} \int_s^T \left(c_u^{la} + \int_0^s \theta_{u,v}^{la} \, dW_v\right) du\,. \quad (3.33)$$

The processes θ_u^{ga} and θ_u^{la} as well as the constants c^{ga} and c^{la} are given through the martingale representations (3.21) for the respective choice of Y.

Proof. The results are straightforward applications of Proposition 3.2.7 with the positive, bounded and $\mathbb{F}$-adapted processes Y given by $Y_t = 1 - G_t$, $t \in [0,T]$, for the gratification annuity and $Y_t = 1$, $t \in [0,T]$, for the simple life annuity. $\qquad\square$

Finally, we turn to the longevity bond. By martingale representation for each $u \in [0,T]$ there exists a predictable process $(\xi_{u,s})_{s \in [0,T]}$, with $\xi_{u,s} = 0$ for $s > u$, such that

$$\mathbb{E}\left[e^{-ru} G_u \mid \mathcal{F}_t\right] = \mathbb{E}\left[e^{-ru} e^{-\Gamma_u} \mid \mathcal{F}_t\right] = k_u + \int_0^{u \wedge t} \xi_{u,s} \, dW_s$$

$$= k_u + \int_0^t \xi_{u,s} \mathbb{1}_{[0,u]}(s) \, dW_s\,, \quad (3.34)$$

with $k_u = \mathbb{E}\left[e^{-ru} G_u\right]$. We set $c := \int_0^T k_u \, du$.

Proposition 3.2.11. *The GKW-decomposition of the discounted intrinsic value process $\widehat{U}^{lb} = (\widehat{U}_t^{lb})_{t \in [0,T]}$ of the longevity bond with respect to W is*

$$\widehat{U}_t^{lb} = \mathbb{E}\left[\int_0^T e^{-ru} G_u \, du \,\Big|\, \mathcal{G}_t\right] = c + \int_0^t \xi_s \, dW_s\,,$$

where the predictable integrand ξ is given as

$$\xi_s = \int_s^T \xi_{u,s}\, du. \tag{3.35}$$

Proof. The discounted survival probability $e^{-ru}G_u$ is bounded and $\mathcal{F}_u$-measurable for every $u \in [0,T]$. Due to Hypothesis (H) we then get

$$\mathbb{E}\left[e^{-ru}G_u \mid \mathcal{G}_t\right] = \mathbb{E}\left[e^{-ru}G_u \mid \mathcal{F}_t\right] = k_u + \int_0^{u \wedge t} \xi_{u,s}\, dW_s$$

$$= k_u + \int_0^t \xi_{u,s} \mathbb{1}_{[0,u]}(s)\, dW_s.$$

Since G is bounded by one, we have that the $\left(\int_0^t \xi_{u,s} \mathbb{1}_{[0,u]}(s) dW_s\right)_{t \in [0,T]}$ are bounded martingales, uniformly in u. We can again apply the stochastic Fubini theorem, as given e.g. in Protter [80, Theorem IV.65], to obtain

$$\mathbb{E}\left[\int_0^T e^{-ru}G_u\, du \,\middle|\, \mathcal{G}_t\right] = \int_0^T \mathbb{E}\left[e^{-ru}G_u \mid \mathcal{G}_t\right] du$$

$$= c + \int_0^t \xi_s\, dW_s\,,$$

where the predictable integrand ξ is as desired. $\square$

Summing up, we have found GKW-decompositions for all discounted intrinsic value processes $\widehat{U}_t^C$ of the claim payoffs C with respect to W of the form

$$\widehat{U}_t^C = \mathbb{E}\left[\widehat{C} \mid \mathcal{G}_t\right] = c^C + \int_0^t \epsilon_s^{C,W}\, dW_s + \int_{0+}^t \epsilon_s^{C,M}\, dM_s\,, \tag{3.36}$$

where the integrands $\epsilon^{C,W}$, $\epsilon^{C,M}$ as well as the constant c^C are claim-specific.

Moreover, the discounted intrinsic value process $\widehat{U}^{lb}$ of the longevity bond, which serves as hedging instrument, is given as

$$\widehat{U}_t^{lb} = c + \int_0^t \xi_s\, dW_s. \tag{3.37}$$

In order to apply mean-variance hedging, as introduced in Subsection 1.1.2, our goal is now to find the GKW decomposition (1.16) of every discounted intrinsic value process of each claim with respect to $\widehat{U}^{lb}$ i.e.

$$\mathbb{E}\left[\widehat{C} \mid \mathcal{G}_t\right] = c^C + \int_0^t \vartheta_s^C\, d\widehat{U}_s^{lb} + L_t^C$$

$$= c^C + \int_0^t \vartheta_s^C \xi_s \, dW_s + L_t^C \tag{3.38}$$

with $c^C = \mathbb{E}[\widehat{C}]$ and L^C a square-integrable martingale, strongly orthogonal to $\mathcal{I}^2(\widehat{U}^{lb})$.

Note again that, because $[M, W] = 0$, we have that $\left(\int_{0+}^t \epsilon_s^{C,M} dM_s\right)_{t \in [0,T]}$ is strongly orthogonal to W and, hence, to $\mathcal{I}^2(W)$. By (3.37), we obviously have $\mathcal{I}^2(\widehat{U}^{lb}) \subseteq \mathcal{I}^2(W)$ such that $\left(\int_{0+}^t \epsilon_s^{C,M} dM_s\right)_{t \in [0,T]}$ is also strongly orthogonal to $\mathcal{I}^2(\widehat{U}^{lb})$.

By (3.36) and (3.38) and due to the uniqueness of the GKW-decomposition (3.38), this implies that the integrand ϑ^C is determined uniquely by the equation

$$\vartheta^C \xi = \epsilon^{C,W} . \tag{3.39}$$

Recall that uniqueness of the stochastic processes in the GKW-decomposition is given up to indistinguishability. In particular, we have uniqueness of the stochastic integrands in the decompositions modulo the following equivalence relation: if $\vartheta, \psi \in L^2(\widehat{U}^{lb})$, then

$$\vartheta \sim \psi \quad \text{if} \quad \int_0^\infty (\vartheta_t - \psi_t)^2 \, d[\widehat{U}^{lb}]_t = 0.$$

For a more thorough investigation of GKW-decompositions we refer to Appendix B. Note that if $\xi_t \neq 0$ for all $t \in [0, T]$, we particularly obtain the following theorem.

Theorem 3.2.12. *If $\xi_t \neq 0$ for all $t \in [0, T]$ then the unique mean-variance hedging strategy for*

1) a pure endowment with respect to a longevity bond is given by

$$\varphi_t^{pe} = \frac{\alpha_t^W}{\xi_t} = \frac{L_{t-}\psi_t}{\int_{t+}^T \xi_{u,t} \, du},$$

$$\varphi_t^{0,pe} = c^{pe} + \int_{0+}^t \varphi_s^{pe} \xi_s dW_s - \varphi_t^{pe}\left(c^{lb} + \int_0^t \xi_s dW_s\right), \tag{3.40}$$

where the predictable integrands ψ and $\xi_{u,\cdot}$ are given by (3.11) for $Z = e^{-rT}$ and (3.35).

The corresponding residual risk R_0^{pe} is given as

$$R_0^{pe} = \mathbb{E}\left[\int_{0+}^T L_{s-}^2 \left(c^{pe} + \int_0^s \psi_v \, dW_v\right)^2 (1 - H_s) d\Gamma_s\right]. \tag{3.41}$$

2) a term insurance with respect to a longevity bond is given as

$$\varphi_t^{ti} = \frac{\beta_t^W}{\xi_t} = \frac{L_{t-}\chi_t}{\int_{t+}^T \xi_{u,t}\, du},$$

$$\varphi_t^{0,ti} = c^{ti} + \int_{0+}^t \varphi_s^{ti} \xi_s dW_s - \varphi_t^{ti}\left(c^{lb} + \int_0^t \xi_s dW_s\right), \tag{3.42}$$

where the predictable integrands χ and $\xi_{u,\cdot}$ are given by (3.15) and (3.35).

The corresponding residual risk R_0^{ti} is given as

$$R_0^{ti} = \mathbb{E}\left[\int_{0+}^T \left(e^{-r(s\wedge\tau)} - L_{s-}\left(c^{ti} + \int_0^s \chi_v\, dW_v - \int_0^s e^{-rv}e^{-\Gamma_v}\, d\Gamma_v\right)\right)^2 (1 - H_s)d\Gamma_s\right]. \tag{3.43}$$

3) a general annuity with respect to a longevity bond is given as

$$\varphi_t^Y = \frac{\rho_t^W}{\xi_t} = \frac{L_{t-} \int_t^T \theta_{u,t}^Y\, du}{\int_{t+}^T \xi_{u,s}\, du},$$

$$\varphi_t^{0,Y} = c^Y + \int_{0+}^t \varphi_s^Y \xi_s dW_s - \varphi_t^Y\left(c^{lb} + \int_0^t \xi_s dW_s\right), \tag{3.44}$$

where the predictable integrands $\theta_{u,\cdot}^Y$ and $\xi_{u,\cdot}$ are given in the martingale representations (3.21) and (3.35).

The corresponding residual risk R_0^Y is given as

$$R_0^Y = \mathbb{E}\left[\int_{0+}^T L_{s-}^2 \left(\int_s^T \left(c_u^Y + \int_0^s \theta_{u,v}^Y\, dW_v\right) du\right)^2 (1 - H_s)d\Gamma_s\right]. \tag{3.45}$$

Proof. The mean-variance hedging strategies follow immediately by Theorem 1.1.30, Equation (3.39) and the assumption that $\xi_t \neq 0$ for all $t \in [0, T]$.

According to Theorem 1.1.30, the residual risk R_0^{pe} for a pure endowment is given as

$$R_0^{pe} = \mathbb{E}[(L_T^{pe})^2] = \mathbb{E}\left[\left(\int_{0+}^T \alpha_s^M\, dM_s\right)^2\right] = \mathbb{E}\left[\int_{0+}^T (\alpha_s^M)^2 (1 - H_s)d\Gamma_s\right]$$

$$= \mathbb{E}\left[\int_{0+}^T L_{s-}^2 \left(c^{pe} + \int_0^s \psi_v\, dW_v\right)^2 (1 - H_s)d\Gamma_s\right],$$

where the third equality follows from the Itô isometry and the fact that $\langle M \rangle_s = \Gamma_{\tau\wedge s}$.

The residual risks for term insurance and general annuity are obtained analogously. □

Remark 3.2.13. *As introduced in Section 1.1.2, the concept of mean-variance hedging is motivated and defined for square-integrable T-contingent claims. The mortality claims, introduced in this section can, of course, all be considered as such claims as they are by definition square-integrable and $\mathcal{G}_T$-measurable. However, we have seen in Remark 3.1.5 that particularly term insurance as well as the general annuities are rather defined as cumulative payment processes. Therefore, the interested reader could also interpret the obtained GKW-decompositions as providing the optimal risk-minimizing strategies for the payment processes. Recall that both concepts, mean-variance hedging as well as risk-minimization, are based in our setting on the GKW-decomposition, see Theorem 1.1.30.*

3.3. Mean-Variance Hedging for Life Insurance Portfolios

For an insurance company it is often more important to hedge the risk of a whole insurance portfolio rather than the risk of a single insurance contract. Following ideas of Biffis and Millossovich [26], we extend the results of the previous subsection to hedging strategies for an insurance portfolio.

Let $I^{pe} = \{x_1, ..., x_n\}, I^{ti} = \{y_1, ..., y_m\}, I^Y = \{z_1, ..., z_k\}$ denote the set of insured persons having purchased coverage through pure endowment, term insurance, and/or general life annuity, respectively. For either of those sets we consider a finite counting measure ϱ^{pe}, ϱ^{ti}, ϱ^Y on $(I^{pe}, \mathcal{P}(I^{pe})), (I^{ti}, \mathcal{P}(I^{ti})), (I^Y, \mathcal{P}(I^Y))$, respectively, allowing the insurance company to weight the risk exposures of the different insured persons to the overall portfolio risks differently.

For every $x \in I^{\cdot}$, we consider its random time of decease τ^x with distribution driven by the continuous increasing and $\mathbb{F}$-adapted hazard process $(\Gamma^x_t)_{t \in [0,T]}$, as introduced in Section 3.2. We write $H^x_t = \mathbb{1}_{\{\tau^x \leq t\}}$, $G^x_t = \mathbb{P}(\tau^x > t \mid \mathcal{F}_t) = e^{-\Gamma^x_t}$ as well as $L^x_t = (1 - H^x_t)e^{\Gamma^x_t}$ and $M^x_t = H^x_t - \int_0^t (1 - H^x_s)d\Gamma^x_s$.

Of course, the insurance company is aware of each single life status $x \in I^{\cdot}$ of its portfolios and we have to expand the filtration setting of our probability space. Denoting by $\mathbb{F}^{H^x}$ the natural filtration generated by the processes $(H^x_t)_{t \in [0,T]}$, we assume the insurance company's complete portfolio information to be represented by the filtrations $\mathbb{G}^{\cdot} = \mathbb{F} \vee \bigvee_{x \in I^{\cdot}} \mathbb{F}^{H^x}$. In this context, we extend the martingale invariance property (Hypothesis (H)) to the filtrations $\mathbb{G}^{\cdot}$, i.e. we assume every $\mathbb{F}$-local martingale to be also a $\mathbb{G}^{\cdot}$-local martingale.

By $\widehat{C}^{\cdot, x}$ we denote the single life discounted payoffs of pure endowment, term insurance and general annuity, associated with $x \in I^{\cdot}$. The weighted, discounted portfolio payoffs $\widehat{C}^{P,pe}$, $\widehat{C}^{P,ti}$ and $\widehat{C}^{P,Y}$ up to time T are then given as

$$\widehat{C}^{P,pe} = \sum_{i=1}^{n} \widehat{C}^{pe,x_i} \varrho^{pe}(x_i) = \sum_{i=1}^{n} e^{-rT}(1 - H^{x_i}_T) \varrho^{pe}(x_i),$$

$$\widehat{C}^{P,ti} = \sum_{j=1}^{m} \widehat{C}^{ti,y_j} \varrho^{ti}(y_j) = \sum_{j=1}^{m} e^{-r\tau^{y_j}} H^{y_j}_T \varrho^{ti}(y_j),$$

$$\widehat{C}^{P,Y} = \sum_{l=1}^{k} \widehat{C}^{Y,z_l} \varrho^Y(Z_l) = \sum_{l=1}^{k} \int_0^T e^{-ru}(1 - H_u^{z_l})Y_u^{z_l}\, du\, \varrho^Y(z_l)\,.$$

In order to apply the results of Section 3.2 for a single life status to the weighted, discounted portfolio payoffs, we assume the following conditional independence relation.

Assumption 3.3.1. *The family* $(\tau^x)_{x \in I^{\cdot}}$ *is conditionally independent given* $\mathcal{F}_T$.

For every $x \in I^{\cdot}$ we denote by ψ^x, χ^x and θ_u^x the predictable integrands of the respective martingale representations (3.11) for $Z = e^{-rT}$, (3.15), and (3.21), related to x. Analogously we write $c^{x,pe}$, $c^{x,ti}$, $c^{x,Y}$, $\alpha^{x,W}$, $\alpha^{x,M}$, $\beta^{x,W}$, $\beta^{x,M}$, $\rho^{x,W}$ and $\rho^{x,M}$ for the constants and integrands in the GKW-decompositions (3.13), (3.16) and (3.22), related to x.

Now we are ready to provide the GKW-decompositions of the discounted intrinsic value processes $\widehat{U}^{P,\cdot}$ of the weighted, portfolio payoffs with respect to W in analogy to the previous subsection.

Proposition 3.3.2. *The GKW-decompositions of the discounted intrinsic value processes* $\widehat{U}^{C^{P,\cdot}} = (\widehat{U}_t^{C^{P,\cdot}})_{t \in [0,T]}$ *of the weighted portfolio payoffs of pure endowments, term insurances or general annuities with respect to* W *are given as*

$$\widehat{U}_t^{P,pe} = \mathbb{E}\left[\widehat{C}^{P,pe} \,\Big|\, \mathcal{G}_t^{pe}\right] = c^{P,pe} + \int_0^t \alpha_s^{P,W}\, dW_s + \sum_{i=1}^{n} \int_{0+}^{t} \alpha_s^{x_i,M} \varrho^{pe}(x_i)\, dM_s^{x_i}\,,$$

$$\widehat{U}_t^{P,ti} = \mathbb{E}\left[\widehat{C}^{P,ti} \,\Big|\, \mathcal{G}_t^{ti}\right] = c^{P,ti} + \int_0^t \beta_s^{P,W}\, dW_s + \sum_{j=1}^{m} \int_{0+}^{t} \beta_s^{y_j,M} \varrho^{ti}(y_j)\, dM_s^{y_j}\,,$$

$$\widehat{U}_t^{P,Y} = \mathbb{E}\left[\widehat{C}^{P,Y} \,\Big|\, \mathcal{G}_t^{Y}\right] = c^{P,Y} + \int_0^t \rho_s^{P,W}\, dW_s + \sum_{l=1}^{k} \int_{0+}^{t} \rho_s^{z_l,M} \varrho^{Y}(z_l)\, dM_s^{z_l}\,,$$

where $c^{P,pe} = \sum_{i=1}^{n} c^{x_i,pe}$, $c^{P,ti} = \sum_{j=1}^{m} c^{y_j,ti}$, $c^{P,pe} = \sum_{l=1}^{k} c^{z_l,Y}$ *and the predictable integrands* $\alpha^{P,W}$, $\beta^{P,W}$ *and* $\rho^{P,W}$ *are given as*

$$\alpha_s^{P,W} = \sum_{i=1}^{n} L_{s-}^{x_i} \psi_s^{x_i} \varrho^{pe}(x_i)\,,$$

$$\beta_s^{P,W} = \sum_{j=1}^{m} L_{s-}^{y_j} \chi_s^{y_j} \varrho^{ti}(y_j)\,,$$

$$\rho_s^{P,W} = \sum_{l=1}^{k} L_{s-}^{z_l} \int_s^T \theta_{u,s}^{z_l}\, du \varrho^Y(z_l)\,.$$

Proof. We illustrate the proof only for the weighted portfolio payoff of pure endowments, as the proofs for term insurances and general annuities are identical. We have

$$\mathbb{E}\left[\left.\widehat{C}^{P,pe}\,\right|\,\mathcal{G}_t^{pe}\right] = \mathbb{E}\left[\left.\sum_{i=1}^n \widehat{C}^{x_i,pe}\,\varrho^{pe}(x_i)\,\right|\,\mathcal{G}_t^{pe}\right] = \sum_{i=1}^n \mathbb{E}\left[\left.\widehat{C}^{x_i,pe}\,\right|\,\mathcal{G}_t^{pe}\right]\varrho^{pe}(x_i)$$

$$= \sum_{i=1}^n \mathbb{E}\left[\left.\widehat{C}^{x_i,pe}\,\right|\,\mathcal{F}_t \vee \mathcal{F}_t^{H^{x_i}}\right]\varrho^{pe}(x_i)$$

$$= \sum_{i=1}^n \left(c^{x_i} + \int_0^t \alpha_s^{x_i,W}\,dW_s + \int_{0+}^t \alpha_s^{x_i,M}\,dM_s^{x_i}\right)\varrho^{pe}(x_i)$$

$$= c^{P,pe} + \int_0^t \sum_{i=1}^n \alpha_s^{x_i,W}\,\varrho^{pe}(x_i)\,dW_s + \sum_{i=1}^n \int_{0+}^t \alpha_s^{x_i,M}\,dM_s^{x_i}\,\varrho^{pe}(x_i)$$

$$= c^{P,pe} + \int_0^t \alpha_s^{P,W}\,dW_s + \sum_{i=1}^n \int_{0+}^t \alpha_s^{x_i,M}\,dM_s^{x_i}\,\varrho^{pe}(x_i)\,,$$

where the third equality follows by Assumption 3.3.1. Note that as M^x and W are strongly orthogonal for all $x \in I^{pe}$, so are $\left(\sum_{i=1}^n \int_0^t \alpha_s^{x_i,M}\varrho^{pe}(x_i)dM_s^{x_i}\right)_{t\in[0,T]}$ and W. The assertion now follows by Lemma B.3 since W is a square integrable martingale. $\qquad\square$

In analogy to Theorem 3.2.12, we can now provide the mean-variance hedging strategy for the weighted, discounted portfolio payoffs with respect to the longevity bond, if $\xi_t > 0$ for all $t \in [0,T]$.

Theorem 3.3.3. *If $\xi_t \neq 0$ for all $t \in [0,T]$ then the unique mean-variance hedging strategies for*

1) a weighted portfolio of pure endowments with respect to a longevity bond is given as

$$\varphi_t^{P,pe} = \frac{\sum_{i=1}^n L_{t-}^{x_i}\psi_t^{x_i}\varrho^{pe}(x_i)}{\int_t^T \xi_{u,t}\,du},$$

$$\varphi_t^{0,P,pe} = c^{P,pe} + \int_{0+}^t \varphi_s^{P,pe}\xi_s dW_s - \varphi_t^{P,pe}\left(c^{lb} + \int_0^t \xi_s dW_s\right), \qquad (3.46)$$

where for every $x_i \in I^{pe}$, the process ψ^{x_i} is the predictable integrand of the respective martingale representation (3.11) for $Z = e^{-rT}$ and $\xi_{u,\cdot}$ the predictable process of the martingale representation (3.21).

The corresponding residual risk $R_0^{P,pe}$ is given as

$$R_0^{P,pe} = \mathbb{E}\left[\left(\sum_{i=1}^n \int_{0+}^T \alpha_s^{x_i,M}\varrho^{pe}(x_i)\,dM_s^{x_i}\right)^2\right]. \qquad (3.47)$$

2) a weighted portfolio of term insurances with respect to a longevity bond is given as

$$\varphi_t^{P,ti} = \frac{\sum_{j=1}^m L_{t-}^{y_j} \chi_t^{y_j} \varrho^{ti}(y_j)}{\int_t^T \xi_{u,t}\, du},$$

$$\varphi_t^{0,ti} = c^{P,ti} + \int_{0+}^t \varphi_s^{P,ti}\xi_s dW_s - \varphi_t^{P,ti}\left(c^{lb} + \int_0^t \xi_s dW_s\right), \tag{3.48}$$

where for every $y_j \in I^{ti}$, the processes χ^{y_j} is the predictable integrand of the respective martingale representation (3.15) and $\xi_{u,\cdot}$ the predictable process of the martingale representation (3.21).

The corresponding residual risk $R_0^{P,ti}$ is given as

$$R_0^{P,ti} = \mathbb{E}\left[\left(\sum_{j=1}^m \int_{0+}^t \beta_s^{y_j,M} \varrho^{ti}(y_j)\, dM_s^{y_j}\right)^2\right]. \tag{3.49}$$

3) a weighted portfolio of general annuities with respect to a longevity bond is given as

$$\varphi_t^{P,Y} = \frac{\sum_{l=1}^k L_{t-}^{z_l} \int_t^T \theta_{u,t}^{Y,z_l}\, du \varrho^Y(z_l)}{\int_t^T \xi_{u,t}\, du},$$

$$\varphi_t^{0,Y} = c^{P,Y} + \int_{0+}^t \varphi_s^{P,Y}\xi_s dW_s - \varphi_t^{P,Y}\left(c^{lb} + \int_0^t \xi_s dW_s\right), \tag{3.50}$$

where for every $z_l \in I^Y$, the processes $\theta_{u,\cdot}^{Y,z_l}$ is the predictable integrand of the respective martingale representation (3.21) and $\xi_{u,\cdot}$ the predictable process of the martingale representation (3.21).

The corresponding residual risk $R_0^{P,Y}$ is given as

$$R_0^{P,Y} = \mathbb{E}\left[\left(\sum_{l=1}^k \int_{0+}^t \rho_s^{z_l,M} \varrho^Y(z_l)\, dM_s^{z_l}\right)^2\right]. \tag{3.51}$$

Proof. The proof is straightforward. □

3.4. Affine Specification of the Mortality Intensity

In this section we assume the hazard process Γ to admit a density, i.e. to be of the form $\Gamma_t = \int_0^t \mu_s\, ds$. The stochastic intensity process $\mu = (\mu_t)_{t \in [0,T]}$ is then assumed to be $\mathbb{F}$-progressively measurable, non-negative and affine.

Moreover, we assume

$$C := \sup_{u \in [0,T]} \mathbb{E}\left[\mu_u^2\right] < \infty . \tag{3.52}$$

Within an affine specification for μ, the derivation of the hedging strategies then boils down to solving well known Riccati ODEs.

In more detail, let μ follow the dynamics

$$\begin{cases} d\mu_t = \delta(t, \mu_t)dt + \sigma(t, \mu_t)dW_t \\ \mu_0 = \overline{\mu} \end{cases} \tag{3.53}$$

for some $\overline{\mu} > 0$, where the drift function δ as well as the instantaneous variance function σ^2 are assumed to have affine dependence on μ, i.e.

$$\begin{aligned} \delta(t, \mu_t) &= d_0(t) + d_1(t)\mu_t , \\ \sigma^2(t, \mu_t) &= v_0(t) + v_1(t)\mu_t , \end{aligned} \tag{3.54}$$

with the deterministic functions d_0, d_1, v_0, and v_1 being bounded and continuous.

It is then a well known fact, see Biffis [25], that for $u \in (t, T]$ we have

$$\mathbb{E}\left[e^{-\int_t^u \mu_s\, ds} \;\middle|\; \mathcal{F}_t \right] = e^{\alpha_u(t) + \beta_u(t)\mu_t} ,$$

where the functions α_u and β_u solve the ODEs

$$\begin{cases} \frac{d\beta_u}{dt}(t) = 1 - d_1(t)\beta_u(t) - \frac{1}{2}v_1(t)\beta_u^2(t) \\ \beta_u(u) = 0 , \end{cases}$$
$$\begin{cases} \frac{d\alpha_u}{dt}(t) = -d_0(t)\beta_u(t) - \frac{1}{2}v_0(t)\beta_u^2(t) \\ \alpha_u(u) = 0 . \end{cases} \tag{3.55}$$

Similarly, for $u \in (t, T]$ we have

$$\mathbb{E}\left[e^{-2\int_t^u \mu_s\, ds} \;\middle|\; \mathcal{F}_t \right] = e^{\tilde{\alpha}_u(t) + \tilde{\beta}_u(t)\mu_t} ,$$

where the functions $\tilde{\alpha}_u$ and $\tilde{\beta}_u$ solve the ODEs

$$\begin{cases} \frac{d\tilde{\beta}_u}{dt}(t) = 2 - d_1(t)\tilde{\beta}_u(t) - \frac{1}{2}v_1(t)\tilde{\beta}_u^2(t) \\ \tilde{\beta}_u(u) = 0 , \end{cases}$$
$$\begin{cases} \frac{d\tilde{\alpha}_u}{dt}(t) = -d_0(t)\tilde{\beta}_u(t) - \frac{1}{2}v_0(t)\tilde{\beta}_u^2(t) \\ \tilde{\alpha}_u(u) = 0 . \end{cases}$$

Finally, for $u \in (t, T]$, we have

$$\mathbb{E}\left[e^{-\int_t^u \mu_s \, ds} \mu_u \,\middle|\, \mathcal{F}_t\right] = e^{\alpha_u(t) + \beta_u(t)\mu_t} \left(\hat{\alpha}_u(t) + \hat{\beta}_u(t)\mu_t\right).$$

Here the functions α_u, β_u are again solutions to (3.55) and $\hat{\alpha}_u$ and $\hat{\beta}_u$ are derived by differentiating (3.55) with respect to u and then solve the ODEs

$$\begin{cases} \frac{d\hat{\beta}_u}{dt}(t) = -d_1(t)\hat{\beta}_u(t) - v_1(t)\beta_u(t)\hat{\beta}_u(t) \\ \hat{\beta}_u(u) = 1, \end{cases}$$

$$\begin{cases} \frac{d\hat{\alpha}_u}{dt}(t) = -d_0(t)\hat{\beta}_u(t) - v_0(t)\beta_u(t)\hat{\beta}_u(t) \\ \hat{\alpha}_u(u) = 0. \end{cases}$$

Note that the non-negativity of μ and assumption (3.52) depend on the model parameters. In particular they are satisfied for the Cox-Ingersoll-Ross process. We refer to Duffie et al. [50] for an extensive study of affine models.

Based on this insight, we get for every $u \in (t, T]$

$$\mathbb{E}\left[e^{-ru}e^{-\Gamma_u} \,\middle|\, \mathcal{F}_t\right] = e^{-ru}e^{-\Gamma_t}\mathbb{E}\left[e^{-\int_t^u \mu_s ds} \,\middle|\, \mathcal{F}_t\right] = e^{-ru}e^{-\Gamma_t}e^{\alpha_u(t) + \beta_u(t)\mu_t}$$

$$= e^{-ru}e^{\alpha_u(0) + \beta_u(0)\bar{\mu}} + \int_0^t e^{-ru}e^{-\Gamma_s}e^{\alpha_u(s) + \beta_u(s)\mu_s}\beta_u(s)\sigma(s, \mu_s)dW_s + X_t^3,$$

where

$$X_t^3 = e^{-ru}\int_0^t e^{\alpha_u(s) + \beta_u(s)\mu_s}e^{-\Gamma_s}\left(\partial_s\alpha_u(s) + \mu_s\partial_s\beta_u(s) + \beta_u(s)\delta(s, \mu_s)\right)ds$$

$$+ e^{-ru}\int_0^t e^{\alpha_u(s) + \beta_u(s)\mu_s}e^{-\Gamma_s}\left(\frac{1}{2}\beta_s^2(u)\sigma^2(s, \mu_s) - \mu_s\right)ds$$

is of finite variation and has to vanish as the conditional expectation on the left hand side is a square integrable continuous martingale.

For $u \in [0, t]$ we note that

$$\mathbb{E}\left[e^{-ru}e^{-\Gamma_u} \,\middle|\, \mathcal{F}_t\right] = e^{-ru}e^{-\Gamma_u} = \lim_{v \nearrow u} \mathbb{E}\left[e^{-ru}e^{-\Gamma_u} \,\middle|\, \mathcal{F}_v\right]$$

$$= \lim_{v \nearrow u} e^{-ru}\left(e^{\alpha_u(0) + \beta_u(0)\bar{\mu}} + \int_0^v e^{-\Gamma_s}e^{\alpha_u(s) + \beta_u(s)\mu_s}\beta_u(s)\sigma(s, \mu_s)dW_s\right)$$

$$= e^{-ru}e^{\alpha_u(0) + \beta_u(0)\bar{\mu}} + \int_0^u e^{-ru}e^{-\Gamma_s}e^{\alpha_u(s) + \beta_u(s)\mu_s}\beta_u(s)\sigma(s, \mu_s)dW_s,$$

where we have used the fact that $\mathbb{F}$-martingales are continuous. This shows that for arbitrary $u \in [0, T]$ we have

$$\mathbb{E}\left[e^{-ru}e^{-\Gamma_u} \mid \mathcal{F}_t\right] = e^{-ru}e^{\alpha_u(0)+\beta_u(0)\overline{\mu}} + \int_0^{t\wedge u} e^{-ru}e^{-\Gamma_s}e^{\alpha_u(s)+\beta_u(s)\mu_s}\beta_u(s)\sigma(s,\mu_s)dW_s$$

$$= e^{-ru}e^{\alpha_u(0)+\beta_u(0)\overline{\mu}} + \int_0^t e^{-ru}e^{-\Gamma_s}e^{\alpha_u(s)+\beta_u(s)\mu_s}\beta_u(s)\sigma(s,\mu_s)\mathbb{1}_{[0,u]}(s)dW_s \ .$$

From this we can directly infer the processes ψ, θ_u^{la} and ξ_u of the martingale representations (3.11) (for the special case $Z = e^{-rT}$), (3.21) (for the case of a simple life annuity with $Y_t = 1, t \in [0,T]$) and (3.35), respectively, to be

$$\psi_s = e^{-rT}e^{-\Gamma_s}\sigma(s,\mu_s)e^{\alpha_T(s)+\beta_T(s)\mu_s}\beta_T(s) \tag{3.56}$$

and

$$\theta_{u,s}^{la} = \xi_{u,s} = e^{-ru}e^{-\Gamma_s}\sigma(s,\mu_s)e^{\alpha_u(s)+\beta_u(s)\mu_s}\beta_u(s)\mathbb{1}_{[0,u]}(s) \ . \tag{3.57}$$

Similarly we get for $u \in [0,T]$

$$\mathbb{E}\left[e^{-ru}e^{-\Gamma_u}\mu_u \mid \mathcal{F}_t\right] = \underbrace{e^{-ru}e^{\alpha_u(0)+\beta_u(0)\overline{\mu}}\left(\hat{\alpha}_u(0)+\hat{\beta}_u(0)\overline{\mu}\right)}_{=:c_u}$$

$$+ \int_0^t \underbrace{e^{-ru}e^{-\Gamma_s}e^{\alpha_u(s)+\beta_u(s)\mu_s}\sigma(s,\mu_s)\left(\hat{\beta}_u(s)+\left(\hat{\alpha}_u(s)+\hat{\beta}_u(s)\mu_s\right)\beta_u(s)\right)\mathbb{1}_{[0,u]}(s)}_{=:\eta(u,s)}dW_s \ .$$

$$\tag{3.58}$$

Note that for all $u \in [0,T]$ and all $t \in [0,T]$, we obtain by (3.52) that

$$\mathbb{E}\left[\mathbb{E}\left[e^{-ru}e^{-\Gamma_u}\mu_u \mid \mathcal{F}_t\right]^2\right] \leq \mathbb{E}\left[\mathbb{E}\left[e^{-2ru}e^{-2\Gamma_u}\mu_u^2 \mid \mathcal{F}_t\right]\right]$$

$$= \mathbb{E}\left[e^{-2ru}e^{-2\Gamma_u}\mu_u^2\right] \leq \mathbb{E}\left[\mu_u^2\right] \leq C \ .$$

Hence, $\left(\int_0^t \eta(u,s)dW_s\right)_{t\in[0,T]}$ is a square-integrable martingale. Furthermore note that due to the Itô isometry and Fubini's theorem, we have by (3.52) that

$$\mathbb{E}\left[\int_0^T \int_0^T \eta(u,s)^2 du\,ds\right] = \int_0^T \mathbb{E}\left[\int_0^T \eta(u,s)^2 ds\right]du$$

$$= \int_0^T \mathbb{E}\left[\left(\int_0^T \eta(u,s)dW_s\right)^2\right]du$$

$$= \int_0^T \mathbb{E}\left[\left(\mathbb{E}\left[e^{-ru}e^{-\Gamma_u}\mu_u \mid \mathcal{F}_T\right]-c_u\right)^2\right]du$$

$$\leq \int_0^T \mathbb{E}\left[\mu_u^2\right] du$$
$$\leq CT < \infty \,,$$

since $c_u \geq 0$ for all $u \in [0,T]$.

Hence, we may apply the stochastic Fubini theorem, as provided e.g. in Protter [80, Theorem IV.65], to Equation 3.58 and obtain

$$\mathbb{E}\left[\int_0^T e^{-ru}e^{-\Gamma_u} d\Gamma_u \,\bigg|\, \mathcal{F}_t\right] = \int_0^T \mathbb{E}\left[e^{-ru}e^{-\Gamma_u}\mu_u \,\big|\, \mathcal{F}_t\right] du$$

$$= \int_0^T e^{-ru}\left\{e^{\alpha_u(0)+\beta_u(0)\overline{\mu}}\left(\hat{\alpha}_u(0) + \hat{\beta}_u(0)\overline{\mu}\right)\right.$$

$$\left. + \int_0^t e^{-\Gamma_s}e^{\alpha_u(s)+\beta_u(s)\mu_s}\sigma(s,\mu_s)\left(\hat{\beta}_u(s) + \left(\hat{\alpha}_u(s) + \hat{\beta}_u(s)\mu_s\right)\beta_u(s)\right)\mathbb{1}_{[0,u]}(s)dW_s\right\}du$$

$$= \int_0^T e^{-ru}e^{\alpha_u(0)+\beta_u(0)\overline{\mu}}\left(\hat{\alpha}_u(0) + \hat{\beta}_u(0)\overline{\mu}\right)du$$

$$+ \int_0^t\int_s^T e^{-ru}e^{-\Gamma_s}e^{\alpha_u(s)+\beta_u(s)\mu_s}\sigma(s,\mu_s)\left(\hat{\beta}_u(s) + \left(\hat{\alpha}_u(s) + \hat{\beta}_u(s)\mu_s\right)\beta_u(s)\right)dudW_s \,.$$

From this we can infer the process χ of the martingale representation (3.15) to equal

$$\chi_s = e^{-\Gamma_s}\sigma(s,\mu_s)\int_s^T e^{-ru}e^{\alpha_u(s)+\beta_u(s)\mu_s}\left(\hat{\beta}_u(s) + \left(\hat{\alpha}_u(s) + \hat{\beta}_u(s)\mu_s\right)\beta_u(s)\right)du \,. \qquad (3.59)$$

Finally, we have for $u \in (t,T]$:

$$\mathbb{E}\left[e^{-ru}\left(1 - G_u\right)e^{-\Gamma_u} \,\big|\, \mathcal{F}_t\right] = e^{-ru}e^{-\Gamma_t}\mathbb{E}\left[e^{-\int_t^u \mu_v dv} \,\big|\, \mathcal{F}_t\right] - e^{-ru}e^{-2\Gamma_t}\mathbb{E}\left[e^{-2\int_t^u \mu_v dv} \,\big|\, \mathcal{F}_t\right]$$

$$= e^{-ru}e^{-\Gamma_t}e^{\alpha_u(t)+\beta_u(t)\mu_t} - e^{-ru}e^{-2\Gamma_t}e^{\tilde{\alpha}_u(t)+\tilde{\beta}_u(t)\mu_t}$$

$$= e^{-ru}\left(e^{\alpha_u(0)+\beta_u(0)\overline{\mu}} - e^{\tilde{\alpha}_u(0)+\tilde{\beta}_u(0)\overline{\mu}}\right)$$

$$+ \int_0^t e^{-ru}e^{-\Gamma_s}\sigma(s,\mu_s)\left(e^{\alpha_u(s)+\beta_u(s)\mu_s}\beta_u(s) - e^{-\Gamma_s}e^{\tilde{\alpha}_u(s)+\tilde{\beta}_u(s)\mu_s}\tilde{\beta}_u(s)\right)dW_s + X_t^4 \,,$$

where

$$X_t^4 = \int_0^t e^{-\Gamma_s}e^{\alpha_u(s)+\beta_u(s)\mu_s}\left(\partial_s\alpha_u(s) + \mu_s\partial_s\beta_u(s) + \beta_u(s)\delta(s,\mu_s) + \frac{1}{2}\beta_s^2(u)\sigma^2(s,\mu_s) - \mu_s\right)ds$$

$$- \int_0^t e^{-2\Gamma_s}e^{\tilde{\alpha}_u(s)+\tilde{\beta}_u(s)\mu_s}\left(\partial_s\tilde{\alpha}_u(s) + \mu_s\partial_s\tilde{\beta}_u(s) + \tilde{\beta}_u(s)\delta(s,\mu_s) + \frac{1}{2}\tilde{\beta}_u^2(s)\sigma^2(s,\mu_s) - 2\mu_s\right)ds$$

is of finite variation and has to vanish.

For $u \in [0, t]$ we get by the same limit arguments as above

$$e^{-ru}\left(1 - G_u\right)e^{-\Gamma_u} = e^{-ru}\left(e^{\alpha_u(0) + \beta_u(0)\bar{\mu}} - e^{\tilde{\alpha}_u(0) + \tilde{\beta}_u(0)\bar{\mu}}\right)$$
$$+ \int_0^u e^{-ru}e^{-\Gamma_s}\sigma(s, \mu_s)\left(e^{\alpha_u(s) + \beta_u(s)\mu_s}\beta_u(s) - e^{-\Gamma_s}e^{\tilde{\alpha}_u(s) + \tilde{\beta}_u(s)\mu_s}\tilde{\beta}_u(s)\right)dW_s \, .$$

Hence we get for arbitrary $u \in [0, T]$:

$$\mathbb{E}\left[e^{-ru}\left(1 - G_u\right)e^{-\Gamma_u} \mid \mathcal{F}_t\right] = e^{-ru}\left(e^{\alpha_u(0) + \beta_u(0)\bar{\mu}} - e^{\tilde{\alpha}_u(0) + \tilde{\beta}_u(0)\bar{\mu}}\right)$$
$$+ \int_0^t e^{-ru}e^{-\Gamma_s}\sigma(s, \mu_s)\left(e^{\alpha_u(s) + \beta_u(s)\mu_s}\beta_u(s) - e^{-\Gamma_s}e^{\tilde{\alpha}_u(s) + \tilde{\beta}_u(s)\mu_s}\tilde{\beta}_u(s)\right)\mathbb{1}_{[0,u]}(s)dW_s \, ,$$

from which we infer the process θ_u^{ga} in the martingale representation (3.21) (for the special case of a gratification annuity with $Y_t = 1 - G_t$, $t \in [0, T]$) to be

$$\theta_{u,s}^{ga} = e^{-ru}e^{-\Gamma_s}\sigma(s, \mu_s)\left(e^{\alpha_u(s) + \beta_u(s)\mu_s}\beta_u(s) - e^{-\Gamma_s}e^{\tilde{\alpha}_u(s) + \tilde{\beta}_u(s)\mu_s}\tilde{\beta}_u(s)\right)\mathbb{1}_{[0,u]}(s) \, . \quad (3.60)$$

Note that if $\beta_u(t) \neq 0$ for almost all $t, u \in [0, T]$ with $t \leq u$, we have $\xi_t \neq 0$ for almost all $t \in [0, T]$ because of (3.57). This, however is the case for most affine models which are applied frequently in the literature, see also the next section.

Theorem 3.4.1. *Given the affine structure* (3.53) *for the mortality intensity, let* $\beta_u(t) \neq 0$ *for almost all* $t, u \in [0, T]$ *with* $t \leq u$. *Then the unique mean-variance hedging strategy for*

1) a pure endowment with respect to a longevity bond is given as

$$\varphi_t^{pe} = \frac{e^{-rT}L_{t-}e^{\alpha_T(t) + \beta_T(t)\mu_t}\beta_T(t)}{\int_t^T e^{-ru}e^{\alpha_u(t) + \beta_u(t)\mu_t}\beta_u(t)\,du} \, ,$$

$$\varphi_t^{0,pe} = c^{pe} + \int_{0+}^t \varphi_s^{pe}\int_s^T e^{-ru}e^{\alpha_u(s) + \beta_u(s)\mu_s}\beta_u(s)\,du\,dW_s$$
$$- \varphi_t^{pe}\left(c^{lb} + \int_0^t \int_s^T e^{-ru}e^{\alpha_u(s) + \beta_u(s)\mu_s}\beta_u(s)\,du\,dW_s\right) .$$

2) a term insurance with respect to a longevity bond is given as

$$\varphi_t^{ti} = \frac{L_{t-}\int_t^T e^{-ru}e^{\alpha_u(t) + \beta_u(t)\mu_t}\left(\hat{\beta}_u(t) + \left(\hat{\alpha}_u(t) + \hat{\beta}_u(t)\mu_t\right)\beta_u(t)\right)du}{\int_t^T e^{-ru}e^{\alpha_u(t) + \beta_u(t)\mu_t}\beta_u(t)\,du} \, ,$$

$$\varphi_t^{0,ti} = c^{ti} + \int_{0+}^t \varphi_s^{ti}\int_s^T e^{-ru}e^{\alpha_u(s) + \beta_u(s)\mu_s}\beta_u(s)\,du\,dW_s$$
$$- \varphi_t^{ti}\left(c^{lb} + \int_0^t \int_s^T e^{-ru}e^{\alpha_u(s) + \beta_u(s)\mu_s}\beta_u(s)\,du\,dW_s\right) .$$

3) a gratification annuity with respect to a longevity bond is given as

$$\varphi_t^{ga} = \frac{L_{t-} \int_t^T e^{-ru} \left(e^{\alpha_u(t)+\beta_u(t)\mu_t} \beta_u(t) - e^{-\Gamma_t} e^{\tilde\alpha_u(t)+\tilde\beta_u(t)\mu_t} \tilde\beta_u(t) \right) du}{\int_t^T e^{-ru} e^{\alpha_u(t)+\beta_u(t)\mu_t} \beta_u(t) \, du}, \tag{3.61}$$

$$\varphi_t^{0,ga} = c^Y + \int_{0+}^t \varphi_s^{ga} \int_s^T e^{-ru} e^{\alpha_u(s)+\beta_u(s)\mu_s} \beta_u(s) \, du dW_s$$

$$- \varphi_t^{ga}(c^{lb} + \int_0^t \int_s^T e^{-ru} e^{\alpha_u(s)+\beta_u(s)\mu_s} \beta_u(s) \, du dW_s).$$

4) a simple life annuity with respect to a longevity bond is given as

$$\varphi_t^{la} = L_{t-}, \tag{3.62}$$

$$\varphi_t^{0,la} = c^Y + \int_{0+}^t \varphi_s^{la} \int_s^T e^{-ru} e^{\alpha_u(s)+\beta_u(s)\mu_s} \beta_u(s) \, du dW_s$$

$$- \varphi_t^{la}(c^{lb} + \int_0^t \int_s^T e^{-ru} e^{\alpha_u(s)+\beta_u(s)\mu_s} \beta_u(s) \, du dW_s).$$

Proof. This is a straightforward consequence of Theorem 3.2.12, as well as Equalities (3.56), (3.57), (3.59), (3.60). $\qquad\square$

3.5. Risk Study for Life Annuities

In this section we perform a risk study for gratification and simple life annuities. Based on numerical simulations, we first compare exemplary paths of the optimal mean-variance hedging strategies as well as surfaces for their residual hedging error. Then we compare the systematic risk parts of both annuities. Remember that both annuities are general annuities according to (3.2), with $Y_t = 1 - G_t$ and $Y_t = 1$, $t \in [0, T]$, respectively.

As in the previous section, we assume the hazard process to be absolutely continuous with respect to the Lebesgue measure and the implied intensity to follow an affine process. There exist several works which estimate different types of affine processes to existing life tables, see e.g. Biffis [25], Dahl and Møller [40], or Luciano and Vigna [69]. For our risk study we particularly focus on affine mortality intensities, following a non-mean-reverting Ornstein-Uhlenbeck process and a non-mean-reverting Feller process, respectively. Both processes are introduced and suggested to be suitable for mortality intensities in Luciano and Vigna [69]. More explicitly, for the Ornstein-Uhlenbeck process we set the drift parameters in (3.54) to $d_0(t) = v_1(t) = 0$ as well as $d_1(t) = d_1 \in \mathbb{R}$, $v_0(t) = v_0 \in \mathbb{R}$, $t \in [0, T]$, and for the Feller process $d_0(t) = v_0(t) = 0$ as well as $d_1(t) = d_1 \in \mathbb{R}$, $v_1(t) = v_1 \in \mathbb{R}$, $t \in [0, T]$. In both cases, we find explicit solutions of the Riccati-ODEs, given in the previous section.

For the non-mean-reverting Ornstein-Uhlenbeck process, we get for $t \in [0, u]$:

$$\beta_u(t) = \frac{1}{d_1}\left(1 - e^{d_1(u-t)}\right), \qquad \alpha_u(t) = \frac{v_0(3 + 2d_1(u-t) + e^{2d_1(u-t)} - 4e^{d_1(u-t)})}{4d_1^3},$$

$$\tilde{\beta}_u(t) = \frac{2}{d_1}\left(1 - e^{d_1(u-t)}\right), \qquad \tilde{\alpha}_u(t) = \frac{v_0(3 + 2d_1(u-t) + e^{2d_1(u-t)} - 4e^{d_1(u-t)})}{d_1^3},$$

$$\hat{\beta}_u(t) = e^{d_1(u-t)}, \qquad \hat{\alpha}_u(t) = \frac{v_0(2e^{d_1(u-t)} - e^{2d_1(u-t)} - 1)}{2d_1^2}.$$

For the non-mean-reverting Feller process, we get for $t \in [0, u]$:

$$\beta_u(t) = \frac{2(e^{\gamma(u-t)} - 1)}{(d_1 - \gamma)(e^{\gamma(u-t)} - 1) - 2\gamma}, \qquad \alpha_u(t) = 0,$$

$$\tilde{\beta}_u(t) = \frac{4(e^{\tilde{\gamma}(u-t)} - 1)}{(d_1 - \tilde{\gamma})(e^{\tilde{\gamma}(u-t)} - 1) - 2\tilde{\gamma}}, \qquad \tilde{\alpha}_u(t) = 0,$$

$$\hat{\beta}_u(t) = \frac{4\gamma^2 e^{\gamma(u-t)}}{((\gamma - d_1)(e^{\gamma(u-t)} - 1) + 2\gamma)^2}, \qquad \hat{\alpha}_u(t) = 0,$$

where $\gamma = \sqrt{d_1^2 + 2v_1}$ and $\tilde{\gamma} = \sqrt{d_1^2 + 4v_1}$.

Note that with the lack of the mean-reversion property, both processes, in contrast to their mean-reverting analogues (the Vasicek and the Cox-Ingersoll-Ross model), are of exponential structure as is illustrated in Figure 3.1. Here and for the following illustrations, the parameters are taken from Luciano and Vigna [69]. Furthermore note that the non-mean-reverting Ornstein-Uhlenbeck process does a priori not show the property of non-negativity. However, with an appropriate choice of the model parameters one can set the probability that the process reaches negative values very small. In particular this is true for the parameters found in Luciano and Vigna [69]. We also respect this issue when we vary some of the model parameters for the illustrations. This way, we still consider the non-mean-reverting Ornstein-Uhlenbeck process as suitable for our results, a common assumption in the literature, see e.g. Schrager [83] or Luciano and Vigna [69].

Based on the simulated paths of the mortality intensity and the affine model parameters, we have numerically generated the optimal mean-variance hedging strategies according to the formulas (3.61) and (3.62), respectively, for the Ornstein-Uhlenbeck and the Feller process. Figures 3.2 and 3.3 show exemplary paths of the strategies for gratification and simple life annuities with maturities $T = 5$ and $T = 30$. Note that the strategies which jump to zero before the maturity show that the insured person died at that time. Hence, the optimal hedging strategies intrinsically offer a reasonable property: if the insured person dies before maturity, there is no further necessity to keep a position in the hedging instrument for this contract.

A remarkable difference between the gratification annuity and the simple life annuity is that for both maturities, the insurance company initially has to go short in the longevity bond in order to hedge the risk exposure of a gratification annuity, whereas it has to go

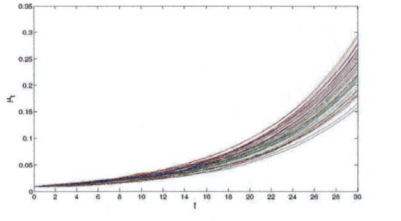

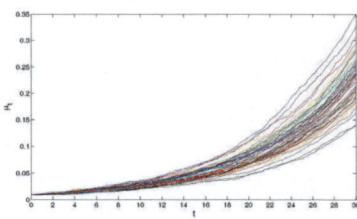

(a) Non-mean-reverting Ornstein-Uhlenbeck process   (b) Non-mean-reverting Feller process

Figure 3.1.: Exemplary paths of a non-mean-reverting Ornstein-Uhlenbeck and a non-mean-reverting Feller process.

long in the longevity bond to hedge the risk exposure of a simple life annuity. This is due to the fact that every rate-payment of the gratification annuity is inferred from the mortality intensity too.

More explicitly, we remark that selling an insurance product means to have a short position in the respective instrument for the insurance company. The rate payments of a single life annuity only depend on the individual survival process $1 - H$, whereas the rate payments of a gratification annuity depend on both the individual survival process $1 - H$ and the mortality rate $1 - G$. The short position in the life annuity therefore yields high overall rate payments with a high realized survival of the insured person. That is why the insurance company has to go long in a longevity bond, as this means to receive higher rate payments with a higher survival rate in the reference portfolio of the longevity index, which can be assumed to be a good proxy for the realized survival process of the insured person. On the contrary, the short position in a gratification annuity means to suffer from both a lower survival rate of the reference portfolio and a higher realized survival of the insured. For a young insured, i.e. at the beginning of the insurance contract, the suffering from a lower survival rate in the reference portfolio dominates the suffering from a higher realized survival of the insured and the insurance company has to go short in a longevity bond in order to cover these rate payments[1]. Only for large maturities and when the insured person gets older, the suffering from his higher realized survival dominates the suffering from lower survival rates in the reference portfolio and a long position in the longevity bond has to cover these long term rate payments.

Another important issue besides the determination of the optimal hedging strategies is the quantification of the residual hedging error. Recall that with the mean-variance hedging approach we have found self-financing trading strategies, which do not perfectly replicate the discounted insurance claims $\widehat{C}$, but yield a value process whose final outcome is optimally close to $\widehat{C}$ in the L^2-norm. However, this value process, although optimal, could still be too far away from the claim and the hedging strategy therefore less reasonable for the insurance company.

---

[1]Note that the longevity bond offers rate payments G.

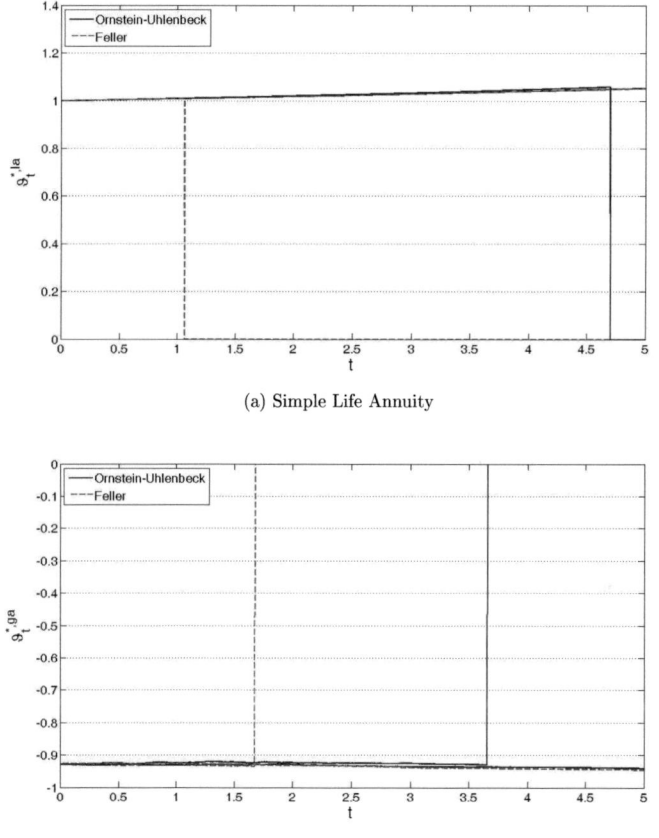

(a) Simple Life Annuity

(b) Gratification Annuity

Figure 3.2.: Exemplary paths of the optimal hedging strategies for a simple life annuity and a gratification annuity with maturity $T = 5$.

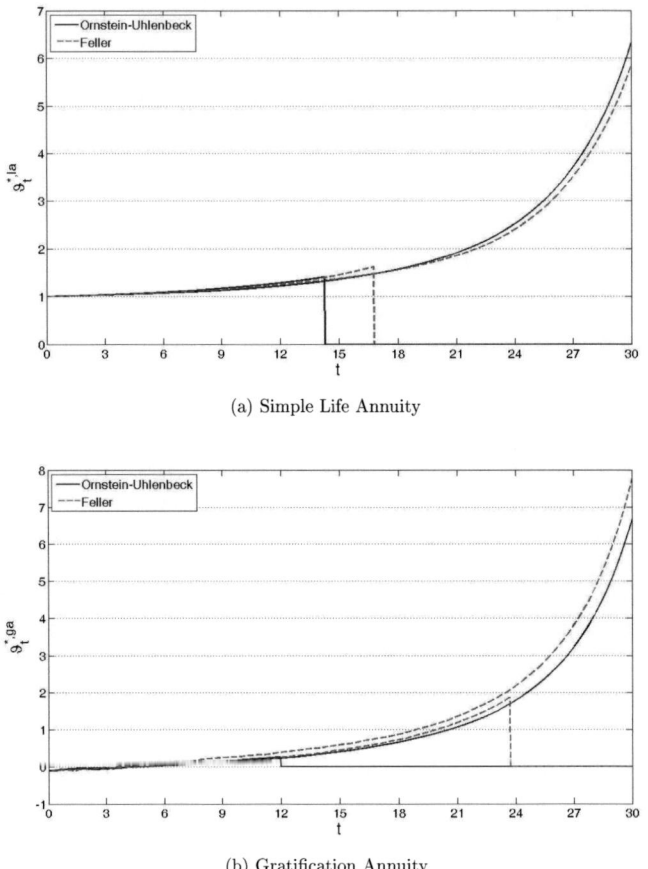

(a) Simple Life Annuity

(b) Gratification Annuity

Figure 3.3.: Exemplary paths of the optimal hedging strategies for a simple life annuity and a gratification annuity with maturity $T = 30$.

Note that by inserting (3.57) or (3.60) into (3.45), we obtain the residual hedging errors R_0^{ga}, R_0^{la} for a gratification and a simple life annuity. Figure 3.4 and Figure 3.5 show numerical results for the residual hedging errors for a gratification annuity and a simple life annuity for different maturities T and different initial mortality intensity levels μ_0 are varying. For both products, the results are again calculated with a mortality intensity following a non-mean-reverting Ornstein-Uhlenbeck process or a non-mean-reverting Feller process, respectively.

For both insurance products, the hedging error increases with increasing maturity, which is not surprising. The remarkable feature, however, is that the residual hedging error of a gratification annuity is considerably lower than the hedging error of a simple life annuity. The levels of R_0 are lower for all considered combinations of maturity and initial mortality intensity under both affine specifications of the mortality intensity. This is due to the fact that the rate payments of the gratification annuity and the longevity bond both depend on the survival rate G, whereas the rate payments of the single life annuity only depend on the individual survival process $1 - H$. Hence, there is a higher correlation between the rate payments of the gratification annuity and the longevity bond than between the rate payments of the single life annuity and the longevity bond. As parts of the mortality risk are forwarded to the insured person through the gratification annuity's rate payments, this yields a good performance of the gratification annuity's optimal hedging scheme. On the contrary, the residual hedging error for existing insurance products like a simple life annuity suggests to consider their optimal hedging strategy rather carefully, especially for longer maturities.

Another point of interest in the context of an insurance claim's risk is the investigation of its systematic and unsystematic parts, see e.g. Norberg [74] and also Section 1.2. The systematic part of an insurance claim's risk can be understood as the part which is due to common risk drivers and its consequences for the insurance company cannot be reduced through diversification. The unsystematic part of an insurance claim's risk can be understood as the part that is due to the insured person's individual characteristics. Its consequences for the insurance company can be reduced through diversification.

Note that in the setting of this chapter, the GKW-decompositions of the different insurance claims intrinsically cover the separation of systematic and unsystematic risk: as we have $\mathbb{G} = \mathbb{F} \vee \mathbb{F}^H$ and $\mathbb{F}$ is generated by W, every claim C can be represented as

$$C = c^C + \underbrace{\int_0^T \epsilon_s^{C,W} \, dW_s}_{\text{systematic risk}} + \underbrace{\int_{0+}^T \epsilon_s^{C,M} \, dM_s}_{\text{unsystematic risk}} \; .$$

As the Brownian motion W is the unique "external" risk driver for all insurance claims, the stochastic integral with respect to W can be considered as the systematic part. The martingales M, however, vary for different insured persons and the integrals with respect to M can therefore be considered as the unsystematic part.

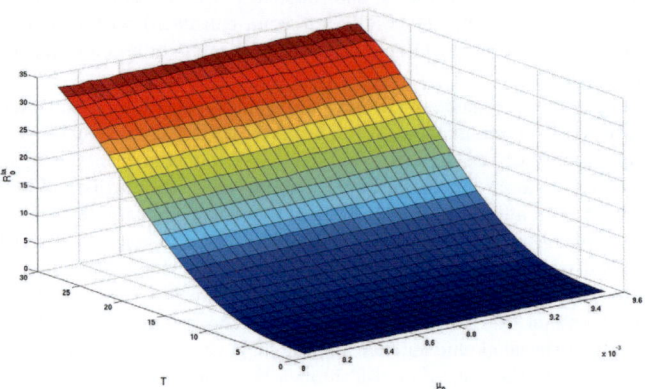

(a) R_0 of a simple life annuity, based on an Ornstein-Uhlenbeck process

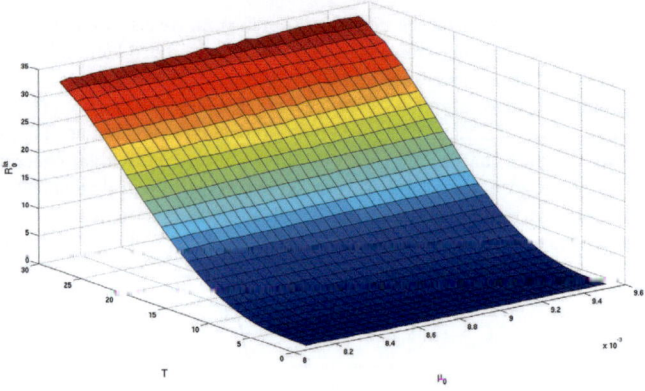

(b) R_0 of a simple life annuity, based on a Feller process

Figure 3.4.: Residual hedging error R_0 for a simple life annuity with mortality intensity, simulated with a non-mean-reverting Ornstein-Uhlenbeck process and a non-mean-reverting Feller process.

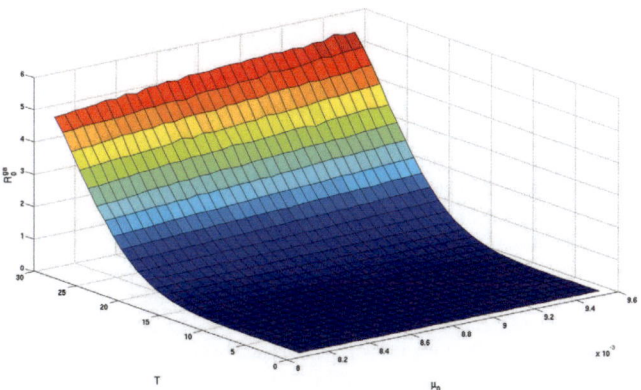

(a) R_0 of a gratification annuity, based on an Ornstein-Uhlenbeck process

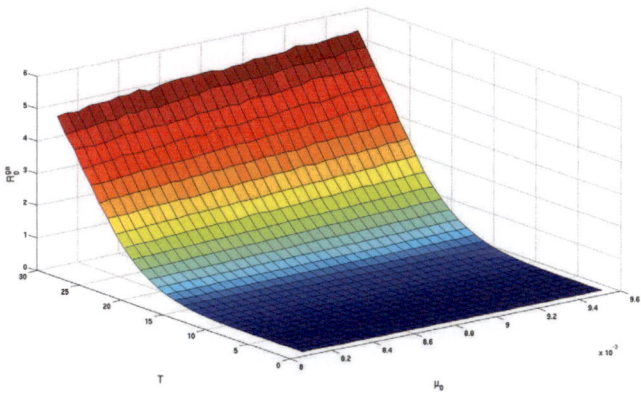

(b) R_0 of a gratification annuity, based on a Feller process

Figure 3.5.: Residual hedging error R_0 for a gratification annuity with mortality intensity, simulated with a non-mean-reverting Ornstein-Uhlenbeck process and a non-mean-reverting Feller process.

As the effects of the unsystematic part diversify through pooling, we now want to compare the systematic risk of a gratification annuity and a simple life annuity. In particular, we can measure the systematic risk SR through

$$SR = \mathbb{E}\left[\left(\int_0^T \epsilon_s^{C,W}\, dW_s\right)^2\right] = \mathbb{E}\left[\int_0^T \left(\epsilon_s^{C,W}\right)^2\, ds\right].$$

In our particular affine framework of this section, Equations (3.57) and (3.60) show that the systematic risk of the simple life annuity is lower or equal than that of the gratification annuity, if $\tilde{\beta}_u(s) \leq 0, \forall s \in [0,T], u \in [s,T]$. This is particularly the case for the mortality intensity μ, following an Ornstein-Uhlenbeck or a Feller process, as well as for most models of practical interest. This is due to the fact that the gratification annuity is exposed to systematic risk in both directions: a structural change in the systematic risk drivers affects both, the insurance company's pool of policyholders and the age cohort from which the rate payments are inferred. A structural decrease in the underlying mortality intensity would e.g. lead to lower claim payments with respect to the insurance portfolio on the one hand, but also to higher annuity rates on the other hand. While a portfolio of simple life annuities would benefit from a structural decrease in the mortality intensity, a portfolio of gratification annuities could also suffer from it. Still, the gratification annuity inherits an advantageous feature from its payout structure: the most common systematic risk exposures of life insurance companies or pension funds are due to increasing longevity. Here, a gratification annuity relaxes the exposure, as increasing longevity leads to lower rates.

The results hence show that the systematic risk of a gratification annuity is higher than that of a simple life annuity, existing on the markets, because the gratification annuity is exposed to risk in any direction. Yet, for the most important systematic risk exposure, increasing longevity, the gratification annuity transfers parts of the systematic risk to the policyholders. For a more thorough investigation of systematic risk in an even more general setting, we refer to Biagini and Schreiber [13].

The investigation of the systematic risk is important under the assumption that no longevity bond is available. With the presence of longevity bonds on the market, however, we have seen in our setting that the complete systematic risk of the insurance claims can be occured. Here, the remaining risk due to hedging errors is considerably lower for the gratification annuity than for a simple life annuity.

Besides the nice "marketing" feature that the insured person gets gratified if he is healthier as originally expected, the gratification annuity therefore shows a better risk behavior than other insurance products already existing on the life insurance market, given the presence of longevity bonds. Moreover we have seen that the insurance company must initially hold a short position in the longevity bond in order to hedge a gratification annuity. This means lower initial overall costs to hedge all longevity products of an insurance company.

All these advantageous features can constitute incentives for insurance companies to introduce gratification annuities as a new life insurance product.

4. Risk-Minimization for General Insurance Contracts

4. Risk-Minimization for General Insurance Contracts

In this chapter we study risk-minimization for a large class of insurance contracts by trading in the assets of the underlying market. Given an underlying $\mathbb{F}$-doubly stochastic Markov chain, we describe different state-dependent types of insurance benefits which cover single payments at maturity, annuity payments and payments at the time of a transition. This definition covers a large set of currently adopted insurance policies.

Given that the $\mathbb{F}$-doubly stochastic Markov chain admits an intensity, we provide the GKW-decompositions for the insurance benefits with respect to a martingale in the reference filtration $\mathbb{F}$. By specifying $\mathbb{F}$ to be the augmented filtration of some N-dimensional Brownian motion $\mathbf{W}$, we then derive risk-minimizing strategies with respect to the assets on an underlying market which is driven by $\mathbf{W}$. Similar to Section 3.4, these results are then further illustrated explicitly within a general affine setting for the intensity.

Large parts of this chapter are based on the findings in Biagini and Widenmann [15].

4.1. General Insurance Contracts

Given the setting of Section 1.1.2, we consider $\mathbb{G} = \mathbb{F}^x \vee \mathbb{F}$ for some reference filtration $\mathbb{F}$, satisfying the usual conditions with $\mathcal{F}_0 = \{\emptyset, \Omega\}$ and some $\mathbb{F}$-doubly stochastic Markov chain X with state-space $\mathcal{K} = \{1, ..., N\}$ as introduced in Appendix A. The following definition of general insurance contracts is based on the definitions of payment processes for rating sensitive claims as given e.g. in Bielecki et al. [23] or Jakubowski and Niewęgłowski [62]. It also covers the concepts of insurance contracts as given in Møller [72] or Norberg [73].

Definition 4.1.1. *A general insurance contract is given by the quadruple $(X; \mathbf{A}; \mathbf{Y}; \mathbf{Z})$, where $X = (X_t)_{t \in [0,T]}$ is the $\mathbb{F}$-doubly stochastic Markov chain, $\mathbf{A} = (A_t^1, ..., A_t^N)_{t \in [0,T]}$ is an N-dimensional $\mathbb{F}$-adapted non-negative increasing process, $\mathbf{Y} = (Y^1, ..., Y^N)$ is an N-dimensional $\mathcal{F}_T$-measurable non-negative random vector, and $\mathbf{Z} = (\mathbf{Z}_t)_{t \in [0,T]}$ with $\mathbf{Z}_t = [Z_t^{j,k}]_{j,k \in \mathcal{K}}$ is an $N \times N$-dimensional $\mathbb{F}$-predictable non-negative process with zeros on the diagonal.*

The different elements of a general insurance contract's quadruple are interpreted as follows. The process X is the insured person's progress in time of sojourning in the states $j \in \mathcal{K}$, considered by the insurance policy. The N-dimensional process $\mathbf{A}$ characterizes the cumulative state-dependent claim payments which are continuously paid up to maturity. The vector $\mathbf{Y}$ characterizes state-dependent "extra" payments at maturity T and the process $\mathbf{Z}$ "immediate" payments at the transition-times from one state to another.

Definition 4.1.2. *The cumulative payment process $D = (D_t)_{t \in [0,T]}$ for every general insurance contract $(X; \mathbf{A}; \mathbf{Y}; \mathbf{Z})$ is given as*

$$D_t := \boldsymbol{Y}^\mathsf{T}\boldsymbol{H}_T\mathbb{1}_{\{t=T\}} + \int_0^t \boldsymbol{H}_s^\mathsf{T}d\boldsymbol{A}_s + \int_{0+}^t (\boldsymbol{Z}_s^\mathsf{T}\boldsymbol{H}_{s-})^\mathsf{T}d\boldsymbol{H}_s$$

$$= \sum_{j=1}^N \left(Y^j H_T^j \mathbb{1}_{\{t=T\}} + \int_0^t H_s^j \, dA_s^j + \sum_{\substack{k=1 \\ k\neq j}}^N \int_{0+}^t Z_s^{j,k} \, dN_s^{jk} \right), \tag{4.1}$$

where the processes $\boldsymbol{H}$ and N^{jk}, $j,k \in \mathcal{K}$, $j \neq k$ are defined by (A.7) and (A.8) in Appendix A.

In the following we work with an arbitrary discounting process S^* as introduced in Definition 1.1.5 which we additionally assume to be $\mathbb{F}$-predictable. In particular, we denote by $(\widehat{\boldsymbol{S}}, 1)$ the discounted market as introduced in Section 1.1. According to Definition 1.1.11, the *discounted* cumulative payment stream $\widehat{D} = (\widehat{D}_t)_{t\in[0,T]}$ is then given as

$$\widehat{D}_t = \frac{\boldsymbol{Y}^\mathsf{T}\boldsymbol{H}_T}{S_T^*}\mathbb{1}_{\{t=T\}} + \int_0^t \frac{1}{S_s^*}\boldsymbol{H}_s^\mathsf{T}d\boldsymbol{A}_s + \int_{0+}^t \frac{1}{S_s^*}(\boldsymbol{Z}_s^\mathsf{T}\boldsymbol{H}_{s-})^\mathsf{T}d\boldsymbol{H}_s$$

$$= \sum_{j=1}^N \left(\frac{Y^j H_T^j}{S_T^*}\mathbb{1}_{\{t=T\}} + \int_0^t \frac{1}{S_s^*} H_s^j dA_s^j + \sum_{\substack{k=1 \\ k\neq j}}^N \int_{0+}^t \frac{1}{S_s^*} Z_s^{j,k} \, dN_s^{jk} \right). \tag{4.2}$$

Remark 4.1.3. *The assumption on $\boldsymbol{A}$ being non-negative and increasing and on $\boldsymbol{Y}$ and $\boldsymbol{Z}$ being non-negative guarantees that D satisfies the properties of cumulative payment processes, introduced in Definition 1.1.8.*

More generally, $\boldsymbol{A}$ could be defined as an N-dimensional $\mathbb{F}$-adapted process of finite variation, $\boldsymbol{Y}$ as an N-dimensional $\mathcal{F}_T$-measurable random vector, and $\boldsymbol{Z}$ as an $N \times N$-dimensional $\mathbb{F}$-predictable process with zeros on the diagonal, see Biagini and Widenmann [15]. In the context of insurance, the payments could then be interpreted to cover also premium payments of the insured person. For example, one could define $\boldsymbol{A}_t = \boldsymbol{C}_t - \boldsymbol{P}_t$, $t \in [0,T]$, with $\boldsymbol{C}_t = (C_t^1, ...C_t^N)^\mathsf{T}$ representing the cumulative state-dependent claim payments (e.g. annuities) and $\boldsymbol{P}_t = (P_t^1, ..., P_t^N)$ the cumulative state-dependent insurance premiums up to maturity. Both processes, $\boldsymbol{P}$ and $\boldsymbol{C}$ are then taken to be $\mathbb{F}$-adapted, càdlàg and increasing.

Provided that the process $\frac{1}{S^*}$ is integrable with respect to $\boldsymbol{A}$ and N^{jk}, $j,k \in \mathcal{K}$, $j \neq k$ and the integral processes satisfy sufficient integrability conditions with respect to $\mathbb{P}$, all results of this chapter also hold for this more general definition.

In the following section, we derive the GKW-decomposition for the discounted cumulative payment process $\widehat{D}$ with respect to some square-integrable $\mathbb{F}$-martingale. Before we continue with these calculations, we first illustrate and connect the definition of a general insurance contract and its cumulative payment process to the insurance contracts, we introduced in Chapters 2 and 3.

Example 4.1.4. *For a pure endowment as introduced in Definition 3.1.4 we consider the state-space of the underlying $\mathbb{F}$-doubly stochastic Markov chain X to be $\mathcal{K} = \{1, 2\}$ with 1 being the state "alive" and 2 the absorbing state "deceased". A pure endowment contract is then a general insurance contract in accordance with Definition 4.1.1 given by the quadruple $(X; \mathbf{0}; (1, 0)^\intercal; \mathbf{0})$, i.e. $\mathbf{A} = \mathbf{0}$, $\mathbf{Z} = \mathbf{0}$ and $\mathbf{Y} = (1, 0)^\intercal$ indicate that there is only a payment of one unit in the case that the insured person is still alive at maturity. The discounted cumulative payment process $\widehat{D}^{pe} = (\widehat{D}_t^{pe})_{t \in [0, T]}$ is then given by*

$$\widehat{D}_t^{pe} = \frac{1}{S_T^*}(1, 0)^\intercal \mathbf{H}_T \mathbb{1}_{\{t=T\}} = \frac{H_T^1}{S_T^*}\mathbb{1}_{\{t=T\}} \quad t \in [0, T].$$

Example 4.1.5. *For a term insurance as introduced in Definition 3.1.4 we again consider the state-space of the underlying $\mathbb{F}$-doubly stochastic Markov chain X to be $\mathcal{K} = \{1, 2\}$ with 1 being the state "alive" and 2 the absorbing state "deceased".*

A term insurance contract is then a general insurance contract given by the quadruple $(X; \mathbf{0}; \mathbf{0}; \mathbf{Z})$ with $\mathbf{Z} := \begin{pmatrix} 0 & 1 \\ 0 & 0 \end{pmatrix}$.

The discounted cumulative payment process $\widehat{D}^{ti} = (\widehat{D}_t^{ti})_{t \in [0, T]}$ is then given by

$$\widehat{D}_t^{ti} = \sum_{j=1}^{2} \sum_{\substack{k=1 \\ k \neq j}}^{2} \int_{0+}^{t} \frac{1}{S_s^*} Z_s^{j,k} \, dN_s^{jk} = \int_{0+}^{t} \frac{1}{S_s^*} \, dN_s^{12} = \frac{1}{S_{\tau_1}^*} \mathbb{1}_{\{\tau_1 \leq t\}} \, ,$$

where τ_1 is the first (and only) jump time of X defined by (A.15).

Example 4.1.6. *For a general life annuity as introduced in Definition 3.1.4 we also consider the state-space of X to be $\mathcal{K} = \{1, 2\}$ with 1 being the state "alive" and 2 the absorbing state "deceased". Then a general annuity contract is given as the quadruple $(X; \left(\left(\int_{0+}^{t} \widetilde{Y}_s ds, 0 \right)^\intercal \right)_{t \in [0, T]}; \mathbf{0}; \mathbf{0})$, where $\widetilde{Y}$ was introduced in Section 3.1 as a positive, bounded, $\mathbb{F}$-adapted stochastic process.*

The discounted cumulative payment process $\widehat{D}^Y = (\widehat{D}_t^Y)_{t \in [0, T]}$ is then given by

$$\widehat{D}_t^{\widetilde{Y}} = \int_0^t \frac{1}{S_s^*} \mathbf{H}_s^\intercal d\mathbf{A}_s = \int_0^t \frac{1}{S_s^*} H_s^1 \widetilde{Y}_s ds \, , \quad t \in [0, T],$$

Example 4.1.7. *We now give a particular example of PPI contracts. In Chapter 2 we investigated PPI contracts covering the risk of unemployment. Here, we consider PPI contracts covering also the events "disabled" and "deceased". Hence, the state space for these PPI products is given as $\mathcal{K} = \{1, 2, 3, 4\}$ with "2" being the state "disabled", "3" the state "unemployed", "4" the absorbing state "deceased", and "1" the state where no insured event is present. For the sake of simplicity, we ignore the presence of waiting, deferment, or requalification periods here.*

A PPI contract is then given as the quadruple

$$(X; \big((0, C_t^2, C_t^3, C_t^4)^\intercal\big)_{t\in[0,T]}; (0, Y^2, Y^3, Y^4)^\intercal; \mathbf{0}).$$

As a PPI contract is usually set up for some payment obligation of the insured person, it guarantees deterministic claim payments $c_1, ..., c_K$ at some a priori given payment dates $0 < T_1 < ... < T_K = T$ in the case of an insured event, see also Section 2.1. Hence, we set $C_t^j = \sum_{k=1}^{K} c_k \mathbf{1}_{[0,t]}(T_k)$, $j = 2, 3, 4$. Moreover, some payment obligations also contain a so-called balloon rate B at the end of the contract which has to be paid on top of the usual installment. If there exists a balloon rate and it is insured, then we set $Y^i = B$, $i = 2, 3, 4$. If there exists no balloon rate or it is not insured, then we set $Y^j = 0$, $j = 2, 3, 4$.

The discounted cumulative payment process $\widehat{D}^{PPI} = (\widehat{D}_t^{PPI})_{t\in[0,T]}$ is then given as

$$\widehat{D}_t^{PPI} = \sum_{j=2}^{4} \left(\frac{Y^j H_T^j}{S_T^*} \mathbf{1}_{\{t=T\}} + \sum_{k=1}^{K} \mathbf{1}_{\{T_k \le t\}} \frac{c_k}{S_{T_k}^*} H_{T_k}^j \right).$$

Remark 4.1.8. *The extra claim payment could also be included in the continuous claim payments. For the sake of distinctness, however, we explicitly separate continuous and extra claim payments. This allows e.g. differentiating more explicitly between annuity type claim payments and single claim payments at maturity as in Example 4.1.7 with balloon rate.*

4.2. GKW-Decomposition for General Insurance Contracts

In this section we provide the GKW-decompositions for general insurance contracts with respect to a square-integrable $\mathbb{F}$-martingale which we will define in Theorem 4.2.3. To this end, we consider a general insurance contract $(X; \mathbf{A}; \mathbf{Y}; \mathbf{Z})$ with discounted payment process $\widehat{D}$, introduced in (4.2), and assume the underlying $\mathbb{F}$-doubly stochastic Markov chain to admit an intensity $\mathbf{\Psi} = \big([\psi_t^{j,k}]_{j,k\in\mathcal{K}}\big)_{t\in[0,T]}$. Moreover, we make the following integrability assumptions.

Assumption 4.2.1. *Given the general insurance contract $(X; \mathbf{A}; \mathbf{Y}; \mathbf{Z})$, let*

$$\mathbb{E}\left[\left(\frac{Y^j}{S_T^*} \right)^2 \right] < \infty, \quad j \in \mathcal{K} \tag{4.3}$$

$$\mathbb{E}\left[\left(\int_0^T \frac{1}{S_u^*} dA_u^j \right)^2 \right] < \infty, \quad j \in \mathcal{K}, \tag{4.4}$$

$$\mathbb{E}\left[\int_{0+}^T \left(\int_{[0,u]} \frac{1}{S_v^*} dA_v^j \right)^2 |\psi_{X_u, j}| du \right] < \infty, \quad j \in \mathcal{K}, \tag{4.5}$$

$$\mathbb{E}\left[\int_{0+}^{T}\left(\frac{Z_u^{j,k}}{S_u^*}\right)^2\psi_{j,k}(u)du\right]<\infty,\quad j,k\in\mathcal{K},j\neq k\,,\tag{4.6}$$

$$\mathbb{E}\left[\left(\int_{0+}^{T}\frac{Z_u^{j,k}}{S_u^*}\psi_{j,k}(u)du\right)^2\right]<\infty,\quad j,k\in\mathcal{K},j\neq k\,.\tag{4.7}$$

Note that (4.3), (4.4), (4.6) and (4.7) ensure that the discounted cumulative payment process $\widehat{D}$, generated by the general insurance contract $(X;\mathbf{A};\mathbf{Y};\mathbf{Z})$, is square integrable.

The following lemma provides a convenient form of the discounted price $\pi_t(D)$ at time $t\in[0,T]$ for the cumulative payment process D after settlement of all payments up to and including time t, as defined in (1.9). We remark that the lemma and its proof are given similarly in Jakubowski and Niewęgłowski [62, Theorem 16.38]. The authors there, however assume the local martingale $\mathbf{M}$ to be a square integrable martingale and the processes $\mathbf{A}$ and $\mathbf{Z}$ to be bounded. Here, we generalize their proof to the case where $\mathbf{A}$ and $\mathbf{Z}$ satisfy the conditions of Assumption 4.2.1.

For notational convenience, we introduce the process $(\boldsymbol{\Delta}_t)_{t\in[0,T]}=\left(\left(\Delta_t^1,...,\Delta_t^N\right)^{\mathsf{T}}\right)_{t\in[0,T]}$ with

$$\Delta_t^j:=[\mathbf{Z}_t\boldsymbol{\Psi}_t^{\mathsf{T}}]_{j,j}=\sum_{\substack{k=1\\k\neq j}}^{N}Z_t^{j,k}\psi_{j,k}(t),\quad j\in\mathcal{K},t\in[0,T]\,.\tag{4.8}$$

Lemma 4.2.2. *Let $(X;\mathbf{A};\mathbf{Y};\mathbf{Z})$ be a general insurance contract satisfying Assumption 4.2.1 then*

$$\begin{aligned}\widehat{\pi}_t(D)&=\mathbb{E}\left[\widehat{D}_T-\widehat{D}_t\,\Big|\,\mathcal{G}_t\right]=\sum_{j=1}^{N}\mathbb{E}\left[\frac{Y^jH_T^j}{S_T^*}+\int_{t+}^{T}\frac{1}{S_u^*}H_u^jdA_u^j+\sum_{\substack{k=1\\k\neq j}}^{N}\int_{t+}^{T}\frac{Z_u^{j,k}}{S_u^*}dN_u^{jk}\,\Bigg|\,\mathcal{G}_t\right]\\&=\sum_{i=1}^{N}H_t^i\sum_{j=1}^{N}\mathbb{E}\left[\frac{Y^jp_{i,j}(t,T)}{S_T^*}+\int_{t+}^{T}\frac{1}{S_u^*}p_{i,j}(t,u)dA_u^j+\sum_{\substack{k=1\\k\neq j}}^{N}\int_{t+}^{T}\frac{Z_u^{j,k}}{S_u^*}p_{i,j}(t,u)\psi_{j,k}(u)du\,\Bigg|\,\mathcal{F}_t\right]\\&=\mathbb{E}\left[\frac{\boldsymbol{P}(t,T)\boldsymbol{Y}}{S_T^*}+\int_{t+}^{T}\frac{\boldsymbol{P}(t,u)}{S_u^*}d\boldsymbol{A}_u+\int_{t+}^{T}\frac{\boldsymbol{P}(t,u)}{S_u^*}\boldsymbol{\Delta}_udu\,\Bigg|\,\mathcal{F}_t\right]^{\mathsf{I}}\boldsymbol{H}_t\,.\end{aligned}\tag{4.9}$$

Proof. We proof the theorem by investigating the different conditional expectations separately.

First note that because $\mathbf{Y}$ is taken to be $\mathcal{F}_T$-measurable and S^* to be $\mathbb{F}$-adapted, we obtain for every $j\in\mathcal{K}$

$$\mathbb{E}\left[\frac{Y^jH_T^j}{S_T^*}\,\Bigg|\,\mathcal{G}_t\right]=\mathbb{E}\left[\frac{Y^j}{S_T^*}\sum_{i=1}^{N}H_t^i\mathbb{E}\left[H_T^j\,|\,\widetilde{\mathcal{G}}_t\right]\,\Bigg|\,\mathcal{G}_t\right]=\mathbb{E}\left[\frac{Y^j}{S_T^*}\sum_{i=1}^{N}H_t^ip_{i,j}(t,T)\,\Bigg|\,\mathcal{G}_t\right]$$

$$= \sum_{i=1}^{N} H_t^i \mathbb{E}\left[\frac{Y^j}{S_T^*}p_{i,j}(t,T)\,\middle|\,\mathcal{F}_t\right],$$

where in the last equality we used the immersion property (A.3) of $\mathbb{F}$-doubly stochastic Markov chains.

Next, by (A.10) of Theorem A.8, we get for $j,k \in \mathcal{K}$, $j \neq k$ that

$$\int_{t+}^{T} \frac{Z_u^{j,k}}{S_u^*}dN_u^{j,k} = \int_{t+}^{T} \frac{Z_u^{j,k}}{S_u^*}dM_u^{j,k} + \int_{t+}^{T} \frac{Z_u^{j,k}}{S_u^*}H_u^j\psi_{j,k}(u)du.$$

Note that because of Remark A.9, as well as (A.13) and (4.6), the integral-process with respect to M^{jk} is a square integrable $\mathbb{G}$-martingale. For every $j,k \in \mathcal{K}$, $j \neq k$, we then obtain with the conditional version of Fubini's theorem, the definition of $\mathbb{F}$-doubly stochastic Markov chains, and Hypothesis (H) as given in (A.3) that

$$\mathbb{E}\left[\int_{t+}^{T}\frac{Z_u^{jk}}{S_u^*}dN_u^{jk}\,\middle|\,\mathcal{G}_t\right] = \mathbb{E}\left[\int_{t+}^{T}\frac{Z_u^{jk}}{S_u^*}H_u^j\psi_{j,k}(u)du\,\middle|\,\mathcal{G}_t\right]$$

$$= \int_{t+}^{T}\mathbb{E}\left[\frac{Z_u^{jk}}{S_u^*}H_u^j\psi_{j,k}(u)\,\middle|\,\mathcal{G}_t\right]du$$

$$= \int_{t+}^{T}\mathbb{E}\left[\mathbb{E}\left[\frac{Z_u^{jk}}{S_u^*}H_u^j\psi_{j,k}(u)\,\middle|\,\widetilde{\mathcal{G}}_t\right]\,\middle|\,\mathcal{G}_t\right]du$$

$$= \int_{t+}^{T}\mathbb{E}\left[\frac{Z_u^{jk}}{S_u^*}\psi_{j,k}(u)\left(\sum_{i=1}^{N}H_t^i p_{i,j}(t,u)\right)\,\middle|\,\mathcal{G}_t\right]du$$

$$= \sum_{i=1}^{N}H_t^i\int_{t+}^{T}\mathbb{E}\left[\frac{Z_u^{jk}}{S_u^*}p_{i,j}(t,u)\psi_{j,k}(u)\,\middle|\,\mathcal{F}_t\right]du$$

$$= \sum_{i=1}^{N}H_t^i\mathbb{E}\left[\int_{t+}^{T}\frac{Z_u^{jk}}{S_u^*}p_{i,j}(t,u)\psi_{j,k}(u)du\,\middle|\,\mathcal{F}_t\right].$$

Finally, for every $j \in \mathcal{K}$ and for fixed $t \in [0,T]$ we define $\widetilde{A}_u^j := \int_{t+}^{u}\frac{1}{S_v^*}dA_v^j$, $u \in [t,T]$. Obviously, $\widetilde{A}^j$ is a finite variation process with $\widetilde{A}_t^j = 0$. By integration by parts as well as by Proposition A.11 we obtain

$$\mathbb{E}\left[\int_{t+}^{T}H_u^j\frac{1}{S_u^*}dA_u^j\,\middle|\,\mathcal{G}_t\right] = \mathbb{E}\left[\int_{t+}^{T}H_u^j d\widetilde{A}_u^j\,\middle|\,\mathcal{G}_t\right]$$

$$= \mathbb{E}\left[\widetilde{A}_T^j H_T^j - \widetilde{A}_t^j H_t^j - \int_{t+}^{T}\widetilde{A}_{u-}^j dH_u^j\,\middle|\,\mathcal{G}_t\right]$$

$$= \mathbb{E}\left[\widetilde{A}_T^j H_T^j - \int_{t+}^{T}\widetilde{A}_{u-}^j dH_u^j\,\middle|\,\mathcal{G}_t\right] = I_1 - I_2,$$

with

$$I_1 := \mathbb{E}[\widetilde{A}_T^j H_T^j \,|\, \mathcal{G}_t], \quad I_2 := \mathbb{E}\left[\int_{t+}^T \widetilde{A}_{u-}^j dH_u^j \,\Big|\, \mathcal{G}_t\right].$$

Since $\widetilde{A}_T$ is $\mathcal{F}_T$-measurable, it follows by Hypothesis (H) that

$$I_1 = \mathbb{E}\left[\widetilde{A}_T^j \mathbb{E}\left[H_T^j \,\Big|\, \widetilde{\mathcal{G}}_t\right] \,\Big|\, \mathcal{G}_t\right] = \sum_{i=1}^K H_t^i \mathbb{E}\left[\widetilde{A}_T^j p_{i,j}(t,v) \,\Big|\, \mathcal{F}_t\right].$$

For I_2 we first note that by (A.9) we have

$$\int_{t+}^T \widetilde{A}_{u-}^j dH_u^j = \int_{t+}^T \widetilde{A}_{u-}^j dM_u^j + \int_{t+}^T \widetilde{A}_{u-}^j \psi_{X_u,j}(u)du.$$

Because of Remark A.9, as well as (A.14) and (4.5), the integral process $\left(\int_t^s \widetilde{A}_{u-}^j dM_u^j\right)_{s\in[t,T]}$ is a square integrable G-martingale. Again by the conditional version of Fubini's theorem, Hypothesis (H) and with the Kolmogorov forward equation (A.6) it then follows that

$$\begin{aligned}
I_2 &= \mathbb{E}\left[\int_{t+}^T \widetilde{A}_{u-}^j dM_u^j + \int_{t+}^T \widetilde{A}_{u-}^j \psi_{X_u,j}(u)du \,\Big|\, \mathcal{G}_t\right]\\
&= \mathbb{E}\left[\int_{t+}^T \widetilde{A}_{u-}^j \sum_{k=1}^K H_u^k \psi_{k,j}(u)du \,\Big|\, \mathcal{G}_t\right] = \int_{t+}^T \mathbb{E}\left[\widetilde{A}_{u-}^j \sum_{k=1}^K H_u^k \psi_{k,j}(u) \,\Big|\, \mathcal{G}_t\right]du\\
&= \int_{t+}^T \mathbb{E}\left[\widetilde{A}_{u-}^j \sum_{k=1}^K \mathbb{E}\left[H_u^k \,\Big|\, \widetilde{\mathcal{G}}_t\right] \psi_{k,j}(u) \,\Big|\, \mathcal{G}_t\right]du\\
&= \sum_{i=1}^K H_t^i \int_{t+}^T \mathbb{E}\left[\widetilde{A}_{u-}^j \left(\sum_{k=1}^K p_{i,k}(t,u)\psi_{k,j}(u)\right) \,\Big|\, \mathcal{F}_t\right]du\\
&= \sum_{i=1}^K H_t^i \mathbb{E}\left[\int_{t+}^T \widetilde{A}_{u-}^j \left(\sum_{k-1}^K p_{i,k}(t,u)\psi_{k,j}(u)\right)du \,\Big|\, \mathcal{F}_t\right]\\
&= \sum_{i=1}^K H_t^i \mathbb{E}\left[\int_{t+}^T \widetilde{A}_{u-}^j dp_{i,j}(t,u) \,\Big|\, \mathcal{F}_t\right].
\end{aligned}$$

Hence, by integration by parts, since $\widetilde{A}^j$ is of finite variation, and since $p(t,\cdot)$ is continuous, we obtain

$$I_1 - I_2 = \sum_{i=1}^K H_t^i \mathbb{E}\left[\widetilde{A}_T p_{i,j}(t,T) - \int_{t+}^T \widetilde{A}_{u-}^j dp_{i,j}(t,u) \,\Big|\, \mathcal{F}_t\right]$$

$$= \sum_{i=1}^{K} H_t^i \mathbb{E}\left[\left. \widetilde{A}_t p_{i,j}(t,t) + \int_{t+}^{T} p_{i,j}(t,u)d\widetilde{A}_u^j \right| \mathcal{F}_t\right]$$

$$= \sum_{i=1}^{K} H_t^i \mathbb{E}\left[\left. \int_{t+}^{T} \frac{1}{S_u^*}p_{i,j}(t,u)dA_u^j \right| \mathcal{F}_t\right],$$

which completes the proof. $\qquad\qquad\square$

In the following theorem we now provide the GKW-decomposition as defined in Definition 1.1.28 for the discounted intrinsic value process of the cumulative payment process $\widehat{D}$, associated to a general insurance contract, with respect to some square-integrable $\mathbb{F}$-martingale.

Theorem 4.2.3. *Let (X, A, Y, Z) be a general insurance contract, satisfying Assumption 4.2.1. Then the GKW-decomposition of the square-integrable discounted intrinsic value process $\widehat{U}^D = (\widehat{U}_t^D)_{t\in[0,T]}$ of D is given as*

$$\widehat{U}_t^D = \widehat{U}_0^D + \int_{0+}^{t} \boldsymbol{\alpha}_u^\intercal d\boldsymbol{m}_u + \int_{0+}^{t} \boldsymbol{\beta}_u^\intercal d\boldsymbol{M}_u, \qquad (4.10)$$

where $\boldsymbol{M}$ is given by (A.9), $\boldsymbol{m} = (\boldsymbol{m}_t)_{t\in[0,T]}$ is a square-integrable $\mathbb{F}$-martingale, given by

$$\boldsymbol{m}_t := \mathbb{E}\left[\left. \frac{\boldsymbol{P}(0,T)\,\boldsymbol{Y}}{S_T^*} + \int_{0}^{T} \frac{\boldsymbol{P}(0,u)}{S_u^*}d\boldsymbol{A}_u + \int_{0+}^{T} \frac{\boldsymbol{P}(0,u)}{S_u^*}\boldsymbol{\Delta}_u du \right| \mathcal{F}_t\right], \qquad (4.11)$$

$\boldsymbol{\alpha}, \boldsymbol{\beta}$ are $\mathbb{R}^N$-valued $\mathbb{G}$-predictable processes defined by

$$\boldsymbol{\alpha}_t = \boldsymbol{L}_{t-} = \boldsymbol{Q}^\intercal(0,t)\boldsymbol{H}_{t-}, \qquad \boldsymbol{\beta}_t = \frac{\boldsymbol{F}(t-,T) + \boldsymbol{Z}_t^\intercal \boldsymbol{H}_{t-}}{S_t^*}, \qquad (4.12)$$

with

$$\boldsymbol{F}(t,T) := S_t^* \mathbb{E}\left[\left. \frac{\boldsymbol{P}(t,T)\,\boldsymbol{Y}}{S_T^*} + \int_{t+}^{T} \frac{\boldsymbol{P}(t,u)}{S_u^*}d\boldsymbol{A}_u + \int_{t+}^{T} \frac{\boldsymbol{P}(t,u)}{S_u^*}\boldsymbol{\Delta}_u du \right| \mathcal{F}_t\right], \qquad (4.13)$$

and $\widehat{U}_0^D = \mathbb{E}[\widehat{D}_T] = \boldsymbol{m}_0^\intercal \boldsymbol{H}_0$.

Remark 4.2.4. *The statement and the proof of Theorem 4.2.3 can be found similarly in Jakubowski and Niewęgłowski [62, Theorem 16.62]. The authors there, however, prove the validity of Decomposition 4.10 only for $t \in [0,T)$. Because we do not have that the integrals on the r.h.s. are continuous, it is a priori not clear if the decomposition also holds for $\widehat{U}_T^D$. Hence, in the following proof we complement the proof of Jakubowski and Niewęgłowski [62] to include the case $t = T$. For the sake of completeness, we also provide the proof for the case $t \in [0,T)$.*

Proof. First note that the processes $\mathbf{m}$ and $\mathbf{F}(\cdot, T)$ are well defined and $\mathbf{m}$ is square-integrable due to Assumption 4.2.1. Moreover, the integrands $\boldsymbol{\alpha}$ and $\boldsymbol{\beta}$ are $\mathbb{G}$-predictable, since Z and S^* are assumed to be $\mathbb{F}$-predictable.

The rest of the proof is separated into two parts. First, we show the validity of the GKW-decomposition for $t = T$. In this case we have

$$
\begin{aligned}
\widehat{U}_T^D = \widehat{D}_T &= \frac{\mathbf{Y}^\intercal \mathbf{H}_T}{S_T^*} + \int_0^T \frac{1}{S_u^*} \mathbf{H}_u^\intercal d\mathbf{A}_u + \int_{0+}^T \frac{1}{S_u^*} (\mathbf{Z}_u^\intercal \mathbf{H}_{u-})^\intercal d\mathbf{H}_u \\
&= \frac{\mathbf{Y}^\intercal \mathbf{H}_T}{S_T^*} + \int_0^T \frac{1}{S_u^*} \mathbf{H}_u^\intercal d\mathbf{A}_u + \int_{0+}^T \frac{1}{S_u^*} (\mathbf{Z}_u^\intercal \mathbf{H}_{u-})^\intercal d\mathbf{M}_u + \int_{0+}^T \frac{1}{S_u^*} (\mathbf{Z}_u^\intercal \mathbf{H}_u)^\intercal \boldsymbol{\Psi}_u^\intercal \mathbf{H}_u du \\
&= \frac{\mathbf{Y}^\intercal \mathbf{H}_T}{S_T^*} + \int_0^T \frac{1}{S_u^*} \mathbf{H}_u^\intercal d\mathbf{A}_u + \int_{0+}^T \frac{1}{S_u^*} (\mathbf{Z}_u^\intercal \mathbf{H}_{u-})^\intercal d\mathbf{M}_u + \int_{0+}^T \frac{1}{S_u^*} \boldsymbol{\Delta}_u^\intercal \mathbf{H}_u du \,, \qquad (4.14)
\end{aligned}
$$

since $\frac{1}{S_u^*} (\mathbf{Z}_u^\intercal \mathbf{H}_u)^\intercal \boldsymbol{\Psi}_u^\intercal \mathbf{H}_u = \frac{1}{S_u^*} \mathbf{H}_u^\intercal \mathbf{Z}_u \boldsymbol{\Psi}_u^\intercal \mathbf{H}_u = \boldsymbol{\Delta}_u^\intercal \mathbf{H}_u$ with notation (4.8). Next, observe that

$$
\begin{aligned}
\frac{1}{S_T^*} \mathbf{Y}^\intercal \mathbf{H}_T &= \frac{1}{S_T^*} (\mathbf{P}(T,T)\mathbf{Y})^\intercal \mathbf{H}_T = \frac{1}{S_T^*} (\mathbf{Q}(0,T)\mathbf{P}(0,T)\mathbf{Y})^\intercal \mathbf{H}_T \\
&= \frac{1}{S_T^*} (\mathbf{P}(0,T)\mathbf{Y})^\intercal \mathbf{Q}(0,T)^\intercal \mathbf{H}_T = \underbrace{\mathbb{E}\left[\frac{1}{S_T^*} \mathbf{P}(0,T)\mathbf{Y} \,\bigg|\, \mathcal{F}_T \right]^\intercal}_{=(\mathbf{m}_T^Y)^\intercal} \mathbf{L}_T \\
&= (\mathbf{m}_0^Y)^\intercal \mathbf{H}_0 + \int_{0+}^T \mathbf{L}_{u-}^\intercal d\mathbf{m}_u^Y + \int_{0+}^T (\mathbf{m}_{u-}^Y)^\intercal d\mathbf{L}_u + \sum_{0 < u \le T} (\Delta \mathbf{m}_u^Y)^\intercal \mathbf{Q}^\intercal(0,u) \Delta \mathbf{H}_u \\
&= (\mathbf{m}_0^Y)^\intercal \mathbf{H}_0 + \int_{0+}^T \mathbf{L}_{u-}^\intercal d\mathbf{m}_u^Y + \int_{0+}^T (\mathbf{m}_{u-}^Y)^\intercal d\mathbf{L}_u \\
&= (\mathbf{m}_0^Y)^\intercal \mathbf{H}_0 + \int_{0+}^T \mathbf{L}_{u-}^\intercal d\mathbf{m}_u^Y + \int_{0+}^T (\mathbf{Q}(0,u)\mathbf{m}_{u-}^Y)^\intercal d\mathbf{M}_u \,,
\end{aligned}
$$

where we set $\mathbf{m}_t^Y := \mathbb{E}\left[\frac{1}{S_T^*} \mathbf{P}(0,T)\mathbf{Y} \,\big|\, \mathcal{F}_t \right]$, $t \in [0,T]$. Note that we used (A.12) in the second equality. We also used the fact that $\mathbf{L}$ is a finite variation process due to (A.11) and that $\mathbf{H}$ does not have simultaneous jumps with the $\mathbb{F}$-martingale $\mathbf{m}^Y$ due to Proposition A.11. Similarly, we obtain

$$
\begin{aligned}
\int_0^T \frac{1}{S_u^*} \mathbf{H}_u^\intercal d\mathbf{A}_u &= \int_0^T \frac{1}{S_u^*} \mathbf{H}_u^\intercal \mathbf{P}(u,u) d\mathbf{A}_u = \int_0^T \frac{1}{S_u^*} \mathbf{H}_u^\intercal \mathbf{Q}(0,u)\mathbf{P}(0,u) d\mathbf{A}_u \\
&= \int_0^T \mathbf{L}_u^\intercal \frac{1}{S_u^*} \mathbf{P}(0,u) d\mathbf{A}_u = \mathbf{L}_T^\intercal \int_0^T \frac{1}{S_u^*} \mathbf{P}(0,u) d\mathbf{A}_u - \int_{0+}^T \int_0^{u-} \frac{1}{S_v^*} \mathbf{P}(0,v) d\mathbf{A}_v d\mathbf{L}_u \\
&= \mathbf{L}_T^\intercal \underbrace{\mathbb{E}\left[\int_0^T \frac{1}{S_u^*} \mathbf{P}(0,u) d\mathbf{A}_u \,\bigg|\, \mathcal{F}_T \right]}_{=\mathbf{m}_T^A} - \int_{0+}^T \int_0^{u-} \frac{1}{S_v^*} \mathbf{P}(0,v) d\mathbf{A}_v d\mathbf{L}_u
\end{aligned}
$$

$$= \mathbf{H}_0^\intercal \mathbf{m}_0^A + \int_{0+}^T (\mathbf{m}_{u-}^A)^\intercal d\mathbf{L}_u + \int_{0+}^T \mathbf{L}_{u-}^\intercal d\mathbf{m}_u^A - \int_{0+}^T \int_0^{u-} \frac{1}{S_v^*} \mathbf{P}(0,v) d\mathbf{A}_v d\mathbf{L}_u$$

$$= \mathbf{H}_0^\intercal \mathbf{m}_0^A + \int_{0+}^T \mathbf{L}_{u-}^\intercal d\mathbf{m}_u^A + \int_{0+}^T \left(\mathbf{Q}(0,u) \mathbb{E}\left[\int_u^T \frac{1}{S_v^*} \mathbf{P}(0,v) d\mathbf{A}_v \;\middle|\; \mathcal{F}_{u-} \right] \right)^\intercal d\mathbf{M}_u ,$$

with $\mathbf{m}_t^A := \mathbb{E}\left[\int_0^T \frac{1}{S_u^*} \mathbf{P}(0,u) d\mathbf{A}_u \;\middle|\; \mathcal{F}_t \right]$, $t \in [0,T]$.

Analogous calculations finally show

$$\int_{0+}^T \frac{1}{S_u^*} \boldsymbol{\Delta}_u^\intercal \mathbf{H}_u du = \mathbf{H}_0^\intercal \mathbf{m}_0^Z + \int_{0+}^T \mathbf{L}_{u-}^\intercal d\mathbf{m}_u^Z$$
$$+ \int_{0+}^T \left(\mathbf{Q}(0,u) \mathbb{E}\left[\int_u^T \frac{1}{S_v^*} \mathbf{P}(0,v) \boldsymbol{\Delta}_v dv \;\middle|\; \mathcal{F}_{u-} \right] \right)^\intercal d\mathbf{M}_u ,$$

with $\mathbf{m}_t^Z := \mathbb{E}\left[\int_{0+}^T \frac{1}{S_u^*} \mathbf{P}(0,u) \boldsymbol{\Delta}_u du \;\middle|\; \mathcal{F}_t \right]$, $t \in [0,T]$. Putting everything together yields

$$\widehat{U}_T^D = \mathbf{H}_0^\intercal \mathbf{m}_0 + \int_{0+}^T \mathbf{L}_{u-}^\intercal d\mathbf{m}_u + \int_{0+}^T \frac{(\mathbf{F}(u-,T) + \mathbf{Z}_u^\intercal \mathbf{H}_{u-})^\intercal}{S_u^*} d\mathbf{M}_u ,$$

with $\mathbf{m} = \mathbf{m}^Y + \mathbf{m}^A + \mathbf{m}^Z$ and $\mathbf{F}(t,T)$ as given in (4.11) and (4.13).

Next, for $t \in [0,T)$ we have

$$\widehat{U}_t^D = I_t^1 + I_t^2 ,$$

with

$$I_t^1 = \int_0^t \frac{\mathbf{H}_u^\intercal}{S_u^*} d\mathbf{A}_u + \int_{0+}^t \frac{(\mathbf{Z}_u^\intercal \mathbf{H}_{u-})^\intercal}{S_u^*} d\mathbf{H}_u$$

and

$$I_t^2 = \mathbb{E}\left[\frac{\mathbf{H}_T^\intercal \mathbf{Y}}{S_T^*} + \int_{t+}^T \frac{\mathbf{H}_u^\intercal}{S_u^*} d\mathbf{A}_u + \int_{t+}^T \frac{(\mathbf{Z}_u^\intercal \mathbf{H}_{u-})^\intercal}{S_u^*} d\mathbf{H}_u \;\middle|\; \mathcal{G}_t \right] .$$

Equivalently to (4.14) we have

$$I_t^1 = \int_0^t \frac{\mathbf{H}_u^\intercal}{S_u^*} d\mathbf{A}_u + \int_{0+}^t \frac{(\mathbf{Z}_u^\intercal \mathbf{H}_{u-})^\intercal}{S_u^*} d\mathbf{M}_u + \int_{0+}^t \frac{(\mathbf{Z}_u^\intercal \mathbf{H}_{u-})^\intercal}{S_u^*} \boldsymbol{\Psi}_u^\intercal \mathbf{H}_u du$$
$$= \int_0^t \frac{\mathbf{H}_u^\intercal}{S_u^*} d\mathbf{A}_u + \int_{0+}^t \frac{(\mathbf{Z}_u^\intercal \mathbf{H}_{u-})^\intercal}{S_u^*} d\mathbf{M}_u + \int_{0+}^t \frac{1}{S_u^*} \boldsymbol{\Delta}_u^\intercal \mathbf{H}_{u-} du .$$

Moreover, with Lemma 4.2.2, it follows that

$$
\begin{aligned}
I_t^2 &= \mathbb{E}\left[\frac{\mathbf{P}(t,T)\mathbf{Y}}{S_T^*} + \int_{t+}^T \frac{\mathbf{P}(t,u)}{S_u^*}d\mathbf{A}_u + \int_{t+}^T \frac{\mathbf{P}(t,u)}{S_u^*}\boldsymbol{\Delta}_u du \,\bigg|\, \mathcal{F}_t\right]^{\mathsf{T}} \mathbf{H}_t \\
&= \mathbb{E}\left[\frac{\mathbf{P}(0,T)\mathbf{Y}}{S_T^*} + \int_{t+}^T \frac{\mathbf{P}(0,u)}{S_u^*}d\mathbf{A}_u + \int_{t+}^T \frac{\mathbf{P}(0,u)}{S_u^*}\boldsymbol{\Delta}_u du \,\bigg|\, \mathcal{F}_t\right]^{\mathsf{T}} \mathbf{Q}^{\mathsf{T}}(0,t)\mathbf{H}_t \\
&= \left(\mathbf{m}_t - \int_0^t \frac{\mathbf{P}(0,u)}{S_u^*}d\mathbf{A}_u - \int_{0+}^t \frac{\mathbf{P}(0,u)}{S_u^*}\boldsymbol{\Delta}_u du\right)^{\mathsf{T}} \mathbf{L}_t \, ,
\end{aligned}
$$

where $\mathbf{m}_t$, $t \in [0,T]$, is defined in (4.11). With integration by parts and the fact that $\mathbf{L}$ is a finite variation process which has no simultaneous jumps with the $\mathbb{F}$-adapted processes $(\mathbf{m}_t)_{t\in[0,T]}$, $\left(\int_0^t \frac{\mathbf{P}(0,u)}{S_u^*}d\mathbf{A}_u\right)_{t\in[0,T]}$ and $\left(\int_0^t \frac{\mathbf{P}(0,u)}{S_u^*}\boldsymbol{\Delta}_u du\right)_{t\in[0,T]}$, it follows that

$$
\begin{aligned}
I_t^2 &= \mathbf{m}_0^{\mathsf{T}}\mathbf{H}_0 + \int_{0+}^t \left(\mathbf{m}_{u-} - \int_0^{u-} \frac{\mathbf{P}(0,v)}{S_v^*}d\mathbf{A}_v - \int_{0+}^{u-} \frac{\mathbf{P}(0,v)}{S_v^*}\boldsymbol{\Delta}_v dv\right)^{\mathsf{T}} dL_u \\
&\quad + \int_{0+}^t \mathbf{L}_{u-}^{\mathsf{T}} \left(d\mathbf{m}_u - \frac{\mathbf{P}(0,u)}{S_u^*}d\mathbf{A}_u - \frac{\mathbf{P}(0,u)}{S_u^*}\boldsymbol{\Delta}_u du\right) \\
&= \mathbf{m}_0\mathbf{H}_0 + \int_{0+}^t \left(\mathbf{Q}(0,u)\left(\mathbf{m}_{u-} - \int_0^{u-} \frac{\mathbf{P}(0,v)}{S_v^*}d\mathbf{A}_v - \int_{0+}^{u-} \frac{\mathbf{P}(0,v)}{S_v^*}\boldsymbol{\Delta}_v dv\right)\right)^{\mathsf{T}} d\mathbf{M}_u \\
&\quad + \int_{0+}^t \mathbf{H}_{u-}^{\mathsf{T}} \left(\mathbf{Q}(0,u)d\mathbf{m}_u - \frac{\mathbf{Q}(0,u)\mathbf{P}(0,u)}{S_u^*}d\mathbf{A}_u - \frac{\mathbf{Q}(0,u)\mathbf{P}(0,u)}{S_u^*}\boldsymbol{\Delta}_u du\right) \\
&= \mathbf{m}_0\mathbf{H}_0 + \int_{0+}^t \left(\mathbf{Q}(0,u)\left(\mathbf{m}_{u-} - \int_0^{u-} \frac{\mathbf{P}(0,v)}{S_v^*}d\mathbf{A}_v - \int_{0+}^{u-} \frac{\mathbf{P}(0,v)}{S_v^*}\boldsymbol{\Delta}_v dv\right)\right)^{\mathsf{T}} d\mathbf{M}_u \\
&\quad + \int_{0+}^t \mathbf{H}_{u-}^{\mathsf{T}} \left(\mathbf{Q}(0,u)d\mathbf{m}_u - \frac{1}{S_u^*}d\mathbf{A}_u - \frac{1}{S_u^*}\boldsymbol{\Delta}_u du\right) \, .
\end{aligned}
$$

Hence we obtain

$$
\begin{aligned}
\widehat{H}_t^D &= I_t^1 + I_t^2 - \mathbf{m}_0^{\mathsf{T}}\mathbf{H}_0 + \int_{0+}^t \mathbf{H}_{u-}^{\mathsf{T}}\mathbf{Q}(0,u)d\mathbf{m}_u + \int_{0+}^t \frac{1}{S_u^*}(\mathbf{Z}_u^{\mathsf{T}}\mathbf{H}_{u-})^{\mathsf{T}}d\mathbf{M}_u \\
&\quad + \int_{0+}^t \left(\mathbf{Q}(0,u)\left(\mathbf{m}_{u-} - \int_0^{u-} \frac{\mathbf{P}(0,v)}{S_v^*}d\mathbf{A}_v - \int_{0+}^{u-} \frac{\mathbf{P}(0,v)}{S_v^*}\boldsymbol{\Delta}_v dv\right)\right)^{\mathsf{T}} d\mathbf{M}_u \\
&= \mathbf{m}_0^{\mathsf{T}}\mathbf{H}_0 + \int_{0+}^t \mathbf{H}_{u-}^{\mathsf{T}}\mathbf{Q}(0,u)d\mathbf{m}_u + \int_{0+}^t \frac{1}{S_u^*}(\mathbf{Z}_u^{\mathsf{T}}\mathbf{H}_{u-} + \mathbf{F}(u-,T))^{\mathsf{T}}d\mathbf{M}_u \, ,
\end{aligned}
$$

where we used the fact that, since $\frac{\mathbf{H}}{S^*}$ and $\mathbf{A}$ have no simultaneous jumps due to Proposition A.11, the integrals $\int_{0+}^t \frac{1}{S_u^*}\mathbf{H}_{u-}d\mathbf{A}_u$ and $\int_{0+}^t \frac{1}{S_u^*}\mathbf{H}_u d\mathbf{A}_u$ are equal for every $t \in [0,T]$. Here, $F(t,T)$ is defined in (4.13).   □

4.3. Risk-Minimization for General Insurance Contracts with Deterministic Payment Structure

In this section we derive risk-minimizing hedging strategies for general insurance contracts with a deterministic payment structure. To this end we need to further specify the hybrid market as given in Chapter 1. First of all, we assume the reference filtration $\mathbb{F} = \mathbb{F}^W$ to be the augmented filtration, generated by some N-dimensional Brownian motion $\mathbf{W}$. For computational reasons, particularly in the affine setting of the next section, we set the dimension N of the Brownian motion equal to the number of states under consideration. We now assume $(\widehat{\mathbf{S}}, 1)$ to be a local $(\mathbb{F}, \mathbb{P})$-martingale, i.e. that the reference filtration generated by the Brownian motion $\mathbf{W}$ particularly covers the full information of the market.

It then follows by the representation theorem with respect to Brownian motion and the non-negativity of $\widehat{\mathbf{S}}$ that there exists a measurable map $\widetilde{\sigma} : [0, T] \times \mathbb{R}^N \to \mathbb{R}^{d \times N}$, such that

$$\widehat{\mathbf{S}}_t = \widehat{\mathbf{S}}_0 + \int_{0+}^t \widetilde{\sigma}(s, \mathbf{S}_s) d\mathbf{W}_s . \tag{4.15}$$

Assumption 4.3.1. *We assume that $\widetilde{\sigma}(t, \mathbf{S}_t)$ is a.s. left-invertible, i.e. that for almost every $(\omega, t) \in \Omega \times [0, T]$ there exists an $\mathbb{F}^W$-predictable $N \times d$-valued matrix $\mathbf{\Gamma}_t(\omega)$ such that $\mathbf{\Gamma}_t \widetilde{\sigma}(t, \mathbf{S}_t) = \mathbf{I}_N$. This particularly implies $N \geq d$.*

From now on we focus on discount factors and insurance contracts with deterministic payment structure.

Assumption 4.3.2. *We assume*

1) $\mathbf{Y}$ is a deterministic vector in $\mathbb{R}^N$.

2) The payment $\mathbf{A} = (\mathbf{A}_t)_{t \in [0, T]}$ is of the form $\mathbf{A}_t = \int_0^t \boldsymbol{\nu}(s) ds$ for some bounded deterministic function $\boldsymbol{\nu} : [0, T] \to \mathbb{R}^N$.

3) $\mathbf{Z} : [0, T] \to \mathbb{R}^{N \times N}$ is a bounded deterministic matrix-valued function.

4) $S^* : [0, T] \to \mathbb{R}$ is a deterministic continuous function.

5) For every $j, k \in \mathcal{K}$, $j \neq k$, $C^{jk} := \sup_{u \in [0, T]} \mathbb{E}\left[\left(\psi_u^{j,k} \right)^2 \right] < \infty.$

Assumption 4.3.2 particularly implies that the integrability conditions of Assumption 4.2.1 hold. Note that the insurance contracts given in Examples 4.1.4, 4.1.5 and 4.1.7 all satisfy 1), 2) and 3) of Assumption 4.3.2. The assumption on S^* being deterministic is applied frequently in the literature, e.g if $\mathbb{P}$ is assumed to be some risk-neutral probability measure and $S_t^* = e^{rt}$ for some constant $r > 0$, see also Chapter 3. More general models in which S^* is considered to be a stochastic process would make the investigation of dependency structures inevitable. Here, this goes beyond the scope of this investigation and we postpone the issue to future research.

Due to the representation theorem with respect to Brownian motion, for every $u \in [0,T]$ and every $i,j \in \mathcal{K}$ there exists some $\boldsymbol{\xi}^{i,j}(u,\cdot) \in L^2(\mathbf{W})$ such that

$$\mathbb{E}\left[p_{i,j}(0,u)\,|\,\mathcal{F}_t\right] = \mathbb{E}\left[p_{i,j}(0,u)\right] + \int_{0+}^{t} \mathbb{1}_{(0,u]}(s)\boldsymbol{\xi}^{i,j}(u,s)^{\mathsf{T}}d\mathbf{W}_s . \tag{4.16}$$

Similarly, because of Assumption 4.3.2 5), for every $u \in [0,T]$ and every $i,j,k \in \mathcal{K}$, $j \neq k$, there exists some $\boldsymbol{\theta}^{i,j,k}(u,\cdot) \in L^2(\mathbf{W})$ such that

$$\mathbb{E}\left[p_{i,j}(0,u)\psi_u^{j,k}\,|\,\mathcal{F}_t\right] = \mathbb{E}\left[p_{i,j}(0,u)\psi_u^{j,k}\right] + \int_{0+}^{t} \mathbb{1}_{(0,u]}(s)\boldsymbol{\theta}^{i,j,k}(u,s)^{\mathsf{T}}d\mathbf{W}_s . \tag{4.17}$$

Theorem 4.3.3. *Given Assumptions 4.3.1 and 4.3.2, the unique 0-admissible risk-minimizing hedging strategy $\bar{\boldsymbol{\xi}} = (\boldsymbol{\xi}, \xi^*)$ for a general insurance claim $(X; \boldsymbol{A}; \boldsymbol{Y}; \boldsymbol{Z})$ with respect to the assets on the underlying market is given as*

$$\boldsymbol{\xi}_t = \sum_{i=1}^{N} L_{t-}^i \sum_{j=1}^{N} \left(\frac{Y_T^j}{S_T^*}\boldsymbol{\xi}^{i,j}(T,t) + \int_t^T \frac{1}{S_u^*}\boldsymbol{\xi}^{i,j}(u,t)\nu_u^j du + \sum_{\substack{k=1\\k\neq j}}^{N} \int_t^T \frac{Z_u^{j,k}}{S_u^*}\boldsymbol{\theta}^{i,j,k}(u,t)du\right)^{\mathsf{T}}\boldsymbol{\Gamma}_t , \tag{4.18}$$

$$\xi_t^* = \widehat{U}_t^D - \widehat{D}_t - \boldsymbol{\xi}_t^{\mathsf{T}}\widehat{\boldsymbol{S}}_t , \tag{4.19}$$

where $\boldsymbol{\Gamma}_t$ is the $\mathbb{F}$-predictable left-inverse of the volatility matrix $\widetilde{\sigma}(t, \widehat{\boldsymbol{S}}_t)$, introduced in Assumption 4.3.1.

Proof. Because of Assumption 4.3.2, the i-th component m^i of the martingale $\mathbf{m}$ in Equation (4.11) is given as

$$m_t^i = \sum_{j=1}^{N}\left(\frac{Y_T^j}{S_T^*}\mathbb{E}\left[p_{i,j}(0,T)\,|\,\mathcal{F}_t\right] + \int_0^T \frac{1}{S_u^*}\mathbb{E}\left[p_{i,j}(0,u)\,|\,\mathcal{F}_t\right]\nu_u^j du\right.$$

$$+ \sum_{\substack{k=1\\k\neq j}}^{N}\int_{0+}^{T} \frac{Z_u^{j,k}}{S_u^*}\mathbb{E}\left[p_{i,j}(0,u)\psi_u^{j,k}\,|\,\mathcal{F}_t\right]du\Bigg)$$

$$= \sum_{j=1}^{N}\left(\frac{Y_T^j}{S_T^*}\mathbb{E}\left[p_{i,j}(0,T)\right] + \int_0^T \frac{1}{S_u^*}\mathbb{E}\left[p_{i,j}(0,u)\right]\nu_u^j du + \sum_{\substack{k=1\\k\neq j}}^{N}\int_{0+}^{T} \frac{Z_u^{j,k}}{S_u^*}\mathbb{E}\left[p_{i,j}(0,u)\psi_u^{j,k}\right]du\right)$$

$$+ \sum_{j=1}^{N}\left(\frac{Y_T^j}{S_T^*}\int_{0+}^{t}\mathbb{1}_{(0,u]}(s)\boldsymbol{\xi}^{i,j}(T,s)^{\mathsf{T}}d\mathbf{W}_s + \int_0^T \frac{1}{S_u^*}\int_{0+}^{t}\mathbb{1}_{(0,u]}(s)\boldsymbol{\xi}^{i,j}(u,s)^{\mathsf{T}}d\mathbf{W}_s\nu_u^j du\right.$$

$$+ \sum_{\substack{k=1\\k\neq j}}^{N}\int_{0+}^{T} \frac{Z_u^{j,k}}{S_u^*}\int_{0+}^{t}\mathbb{1}_{(0,u]}(s)\boldsymbol{\theta}^{i,j,k}(u,s)^{\mathsf{T}}d\mathbf{W}_s du\Bigg) .$$

By Assumption 4.3.2, Fubini's theorem and the Itô isometry it then follows for every $i, j \in \mathcal{K}$ that

$$
\mathbb{E}\left[\int_{0+}^{T}\int_{0+}^{T}\left(\frac{1}{S_u^*}\mathbb{1}_{(0,u]}(s)\|\boldsymbol{\xi}^{i,j}(u,s)\|\nu_u^j\right)^2 duds\right] \leq K_1^2 \int_{0+}^{T}\mathbb{E}\left[\int_{0+}^{T}\|\boldsymbol{\xi}^{i,j}(u,s)\|^2 ds\right] du
$$

$$
= K_1^2 \int_{0+}^{T}\mathbb{E}\left[\left(\int_{0+}^{T}\boldsymbol{\xi}^{i,j}(u,s)^{\mathsf{T}}d\mathbf{W}_s\right)^2\right] du
$$

$$
= K_1^2 \int_{0+}^{T}\mathbb{E}\left[\left(\mathbb{E}\left[p_{i,j}(0,u)\,|\,\mathcal{F}_T\right]-\mathbb{E}\left[p_{i,j}(0,u)\right]\right)^2\right] du
$$

$$
\leq K_1^2 T < \infty
$$

for some constant $K_1 > 0$.

Due to Assumption 4.3.2 5), we similarly have for every $i, j, k \in \mathcal{K}$, $j \neq k$ that

$$
\mathbb{E}\left[\int_{0+}^{T}\int_{0+}^{T}\left(\frac{Z_u^{j,k}}{S_u^*}\mathbb{1}_{(0,u]}(s)\|\boldsymbol{\theta}^{i,j,k}(u,s)\|\right)^2 duds\right] \leq K_2^2 \int_{0+}^{T}\mathbb{E}\left[\int_{0+}^{T}\|\boldsymbol{\theta}^{i,j,k}(u,s)\|^2 ds\right] du
$$

$$
= K_2^2 \int_{0+}^{T}\mathbb{E}\left[\left(\int_{0+}^{T}\boldsymbol{\theta}^{i,j,k}(u,s)^{\mathsf{T}}d\mathbf{W}_s\right)^2\right] du
$$

$$
= K_2^2 \int_{0+}^{T}\mathbb{E}\left[\left(\mathbb{E}\left[p_{i,j}(0,u)\psi_u^{j,k}\,|\,\mathcal{F}_T\right]-\mathbb{E}\left[p_{i,j}(0,u)\right]\right)^2\right] du
$$

$$
\leq K_2^2 \int_{0+}^{T}\mathbb{E}\left[(\psi_u^{j,k})^2\right] du
$$

$$
\leq K_2^2 CT < \infty
$$

for some constant $K_2 > 0$.

Therefore, we can apply the stochastic version of Fubini's theorem, see Protter [80, Theorem IV.65], and obtain

$$
m_t^i = \sum_{j=1}^{N}\left(\frac{Y_T^j}{S_T^*}\mathbb{E}\left[p_{i,j}(0,T)\right]+\int_0^T\frac{1}{S_u^*}\mathbb{E}\left[p_{i,j}(0,u)\right]\nu_u^j du+\sum_{\substack{k=1\\k\neq j}}^{N}\int_{0+}^{T}\frac{Z_u^{j,k}}{S_u^*}\mathbb{E}\left[p_{i,j}(0,u)\psi_u^{j,k}\right]du\right)
$$

$$
+\sum_{j=1}^{N}\int_{0+}^{t}\left(\frac{Y_T^j}{S_T^*}\boldsymbol{\xi}^{i,j}(T,s)+\int_s^T\frac{1}{S_u^*}\boldsymbol{\xi}^{i,j}(u,s)\nu_u^j du+\sum_{\substack{k=1\\k\neq j}}^{N}\int_s^T\frac{Z_u^{j,k}}{S_u^*}\boldsymbol{\theta}^{i,j,k}(u,s)du\right)^{\mathsf{T}}d\mathbf{W}_s\,.
$$

Because of (4.15), this finally implies

$$
dm_t^i = \sum_{j=1}^{N} \Big(\frac{Y_T^j}{S_T^*} \boldsymbol{\xi}^{i,j}(T,t) + \int_t^T \frac{1}{S_u^*} \boldsymbol{\xi}^{i,j}(u,t) \nu_u^j du + \sum_{\substack{k=1 \\ k \neq j}}^{N} \int_t^T \frac{Z_u^{j,k}}{S_u^*} \boldsymbol{\theta}^{i,j,k}(u,t) du \Big)^{\mathsf{T}} d\mathbf{W}_t
$$

$$
= \sum_{j=1}^{N} \Big(\frac{Y_T^j}{S_T^*} \boldsymbol{\xi}^{i,j}(T,t) + \int_t^T \frac{1}{S_u^*} \boldsymbol{\xi}^{i,j}(u,t) \nu_u^j du + \sum_{\substack{k=1 \\ k \neq j}}^{N} \int_t^T \frac{Z_u^{j,k}}{S_u^*} \boldsymbol{\theta}^{i,j,k}(u,t) du \Big)^{\mathsf{T}} \boldsymbol{\Gamma}_t d\widehat{\mathbf{S}}_t \ .
$$

With Theorem 4.2.3, we therefore obtain the GKW-decomposition for the discounted intrinsic value process $\widehat{U}^D$ of D as

$$
\widehat{U}_t^D = \widehat{U}_0^{\widehat{D}} + \int_{0+}^t \Big\{ \sum_{i=1}^N \sum_{j=1}^N L_{s-}^i \Big(\frac{Y_T^j}{S_T^*} \boldsymbol{\xi}^{i,j}(T,t) + \int_t^T \frac{1}{S_u^*} \boldsymbol{\xi}^{i,j}(u,t) \nu_u^j du
$$

$$
+ \sum_{\substack{k=1 \\ k \neq j}}^N \int_t^T \frac{Z_u^{j,k}}{S_u^*} \boldsymbol{\theta}^{i,j,k}(u,t) du \Big)^{\mathsf{T}} \Big\} d\mathbf{W}_s + \int_{0+}^t \boldsymbol{\beta}_s^{\mathsf{T}} d\mathbf{M}_s \qquad (4.20)
$$

$$
= \widehat{U}_0^{\widehat{D}} + \int_{0+}^t \Big\{ \sum_{i=1}^N \sum_{j=1}^N L_{s-}^i \Big(\frac{Y_T^j}{S_T^*} \boldsymbol{\xi}^{i,j}(T,t) + \int_t^T \frac{1}{S_u^*} \boldsymbol{\xi}^{i,j}(u,t) \nu_u^j du
$$

$$
+ \sum_{\substack{k=1 \\ k \neq j}}^N \int_t^T \frac{Z_u^{j,k}}{S_u^*} \boldsymbol{\theta}^{i,j,k}(u,t) du \Big)^{\mathsf{T}} \Big\} \boldsymbol{\Gamma}_s d\widehat{\mathbf{S}}_s + \int_{0+}^t \boldsymbol{\beta}_s^{\mathsf{T}} d\mathbf{M}_s \qquad (4.21)
$$

Because the integrands with respect to the Brownian motion $\mathbf{W}$ and with respect to $\mathbf{M}$ are $\mathbb{G}$-predictable, because $\mathbf{M}$ and $\mathbf{W}$ are strongly orthogonal and because $\mathbf{W}$ is square-integrable, we obtain with Lemma B.3 that the decomposition (4.20) is the GKW-decomposition of $\widehat{U}^D$ with respect to $\mathbf{W}$, i.e. that the integral process $\Big(\int_{0+}^t \boldsymbol{\beta}_s^{\mathsf{T}} d\mathbf{M}_s \Big)_{t \in [0,T]}$ is a square-integrable martingale, strongly orthogonal to $\mathcal{I}^2(\mathbf{W})$. Note however that because of (4.15) and Assumption 4.3.1, we have $\mathcal{I}^2(\widehat{\mathbf{S}}) = \mathcal{I}^2(\mathbf{W})$, such that $\Big(\int_{0+}^t \boldsymbol{\beta}_s^{\mathsf{T}} d\mathbf{M}_s \Big)_{t \in [0,T]}$ is strongly orthogonal to $\mathcal{I}^2(\widehat{\mathbf{S}})$ as well. In particular, the decomposition (4.21) provides the GKW-decomposition (1.16) for $\widehat{U}^D$ with respect to $\widehat{\mathbf{S}}$, which we require to describe the risk-minimizing hedging strategies according to Section 1.1.2.

The result then follows immediately from Theorem 1.1.30. $\qquad\qquad \square$

4.4. Risk-Minimization for General Insurance Contracts within an Affine Specification for the Intensities

In the same setting as in Section 4.3 we now specify the risk-minimizing hedging strategies of the previous exemplary application within a general affine setting for the intensities.

In addition to Assumption 4.3.2, we now assume the following.

Assumption 4.4.1. *1) For every $0 \leq t \leq u \leq T$ and every $j, k \in \mathcal{K}$, $j \neq k$, the entry $p_{j,k}(t,u)$ of the transition matrix $\boldsymbol{P}(t,u)$ is of the form*

$$p_{j,k}(t,u) = 1 - e^{-\int_t^u \psi_v^{j,k} dv} , \qquad (4.22)$$

where $\psi^{j,k}$ are the respective entries of the intensity matrix $\boldsymbol{\Psi}$.

2) For every $u \in [0,T]$ and every $j, k \in \mathcal{K}$, $j \neq k$, $\psi_u^{j,k}$ is of the form

$$\psi_u^{j,k} = (\boldsymbol{b}^{j,k})^\mathsf{T} \boldsymbol{\mu}_u + c^{j,k} , \qquad (4.23)$$

where $\boldsymbol{b}^{j,k} \in \mathbb{R}^N$, $c^{j,k} \in \mathbb{R}$ and $\boldsymbol{\mu} = (\boldsymbol{\mu}_t)_{t \in [0,T]}$ an $\mathbb{R}^N$-valued affine process as specified e.g. in Biffis [25, Section 3 and Appendix A] or Duffie et al. [49]. Here, $\boldsymbol{\mu}$ is a Markov process with respect to $\mathbb{F}^W$, given as the strong solution to the SDE

$$d\boldsymbol{\mu}_t = \boldsymbol{\delta}(t, \boldsymbol{\mu}_t)dt + \boldsymbol{\sigma}(t, \boldsymbol{\mu}_t)d\boldsymbol{W}_t , \qquad (4.24)$$

where for $t \in [0,T]$, $\boldsymbol{x} \in \mathbb{R}^N$ and $i, j \in \{1, ..., N\}$

$$\boldsymbol{\delta}(t, \boldsymbol{x}) = \boldsymbol{d}^0(t) + (\boldsymbol{d}^1(t))^\mathsf{T} \boldsymbol{x} \qquad (4.25)$$

$$[\boldsymbol{\sigma}(t, \boldsymbol{x})\boldsymbol{\sigma}(t, \boldsymbol{x})^\mathsf{T}]_{i,j} = [\boldsymbol{V}^0(t)]_{i,j} + (\boldsymbol{V}^1(t))_{i,j}^\mathsf{T} \boldsymbol{x} , \qquad (4.26)$$

with coefficient functions $\boldsymbol{d}^0, \boldsymbol{d}^1, \boldsymbol{V}^0$ and $\boldsymbol{V}^1$, taking values in $\mathbb{R}^N, \mathbb{R}^{N \times N}, \mathbb{R}^{N \times N}$ and $\mathbb{R}^{N \times N \times N}$, respectively.

3) The process $\boldsymbol{\mu}$ is such that

$$C := \sup_{u \in [0,T]} \mathbb{E}[\|\boldsymbol{\mu}_u\|^2] < \infty . \qquad (4.27)$$

Note that this particularly implies that for every $j, k \in \mathcal{K}$, $j \neq k$, we have

$$\sup_{u \in [0,T]} \mathbb{E}[(\psi_u^{j,k})^2] < \infty . \qquad (4.28)$$

With these assumptions and under some technical conditions, presented in Duffie et al. [49], we obtain for every $0 \leq t \leq u \leq T$ and every $i, j \in \mathcal{K}$ with $i \neq j$ that

$$\mathbb{E}\left[p_{i,j}(t,u)\mid\mathcal{F}_t\right]=\mathbb{E}[1-e^{-\int_t^u\psi_v^{i,j}dv}\mid\mathcal{F}_t]=1-e^{\alpha_u^{i,j}(t)+(\boldsymbol{\beta}_u^{i,j}(t))^\intercal\boldsymbol{\mu}_t}\;,\tag{4.29}$$

$$\mathbb{E}\left[p_{i,i}(t,u)\mid\mathcal{F}_t\right]=\mathbb{E}[1-\sum_{\substack{j=1\\j\neq i}}^N p_{i,j}(t,u)\mid\mathcal{F}_t]=1-\sum_{\substack{j=1\\j\neq i}}^N\mathbb{E}\left[1-e^{-\int_t^u\psi_v^{i,j}dv}\mid\mathcal{F}_t\right]$$

$$=2-N+\sum_{\substack{j=1\\j\neq i}}^N e^{\alpha_u^{i,j}(t)+(\boldsymbol{\beta}_u^{i,j}(t))^\intercal\boldsymbol{\mu}_t}\;.\tag{4.30}$$

Similarly, we obtain for every $0\leq t\leq u\leq T$ and every $i,j,k\in\mathcal{K}$ with $i\neq j,\,j\neq k$

$$\mathbb{E}\left[p_{i,j}(t,u)\psi_u^{j,k}\mid\mathcal{F}_t\right]=\mathbb{E}\left[(1-e^{-\int_t^u\psi_v^{i,j}du})\psi_u^{j,k}\mid\mathcal{F}_t\right]=\mathbb{E}\left[\psi_u^{j,k}\mid\mathcal{F}_t\right]-\mathbb{E}\left[e^{-\int_t^u\psi_v^{i,j}du}\psi_u^{j,k}\mid\mathcal{F}_t\right]$$

$$=e^{\widetilde{\alpha}_u(t)+(\widetilde{\boldsymbol{\beta}}_u(t))^\intercal\boldsymbol{\mu}_t}\big(\widetilde{\alpha}_u^{j,k}(t)+(\widetilde{\boldsymbol{\beta}}_u^{j,k}(t))^\intercal\boldsymbol{\mu}_t\big)$$

$$-e^{\alpha_u^{i,j}(t)+(\boldsymbol{\beta}_u^{i,j}(t))^\intercal\boldsymbol{\mu}_t}\big(\widehat{\alpha}_u^{j,k}(t)+(\widehat{\boldsymbol{\beta}}_u^{j,k}(t))^\intercal\boldsymbol{\mu}_t\big)\;,\tag{4.31}$$

$$\mathbb{E}\left[p_{j,j}(t,u)\psi_u^{j,k}\mid\mathcal{F}_t\right]=(2-N)\mathbb{E}\left[\psi_u^{j,k}\mid\mathcal{F}_t\right]+\sum_{\substack{l=1\\l\neq j}}^N\mathbb{E}\left[e^{-\int_t^u\psi_u^{j,l}du}\psi_u^{j,k}\mid\mathcal{F}_t\right]$$

$$=(2-N)e^{\widetilde{\alpha}_u(t)+(\widetilde{\boldsymbol{\beta}}_u(t))^\intercal\boldsymbol{\mu}_t}\big(\widetilde{\alpha}_u^{j,k}(t)+(\widetilde{\boldsymbol{\beta}}_u^{j,k}(t))^\intercal\boldsymbol{\mu}_t\big)$$

$$+\sum_{\substack{l=1\\l\neq j}}^N e^{\alpha_u^{j,l}(t)+(\boldsymbol{\beta}_u^{j,l}(t))^\intercal\boldsymbol{\mu}_t}\big(\widehat{\alpha}_u^{j,k}(t)+(\widehat{\boldsymbol{\beta}}_u^{j,k}(t))^\intercal\boldsymbol{\mu}_t\big)\;.\tag{4.32}$$

For every $0\leq t\leq u\leq T$ and every combination of $i,j,k,l\in\mathcal{K}$, considered in equations (4.29), (4.30), (4.31) and (4.32), the functions $\alpha_u^{i,j},\,\boldsymbol{\beta}_u^{i,j}$ solve the ODEs

$$\frac{d\boldsymbol{\beta}_u^{i,j}}{dt}(t)=\mathbf{c}^{i,j}-\mathbf{d}^1(t)^\intercal\boldsymbol{\beta}_u^{i,j}(t)-\frac{1}{2}(\boldsymbol{\beta}_u^{i,j}(t))^\intercal\mathbf{V}^1(t)\boldsymbol{\beta}_u^{i,j}(t)\;,$$

$$\frac{d\alpha_u^{i,j}}{dt}(t)=b^{i,j}\quad\mathbf{d}^0(t)^\intercal\boldsymbol{\beta}_u^{i,j}(t)-\frac{1}{2}(\boldsymbol{\beta}_u^{i,j}(t))^\intercal\mathbf{V}^0(t)\boldsymbol{\beta}_u^{i,j}(t)\;,$$

with terminal conditions $\alpha_u^{i,j}(u)=0$ and $\boldsymbol{\beta}_u^{i,j}(u)=0$, whereas the functions $\widetilde{\alpha}_u,\,\widetilde{\boldsymbol{\beta}}_u$ solve the ODEs

$$\frac{d\widetilde{\boldsymbol{\beta}}_u}{dt}(t)=-\mathbf{d}^1(t)^\intercal\widetilde{\boldsymbol{\beta}}_u(t)-\frac{1}{2}(\widetilde{\boldsymbol{\beta}}_u(t))^\intercal\mathbf{V}^1(t)\widetilde{\boldsymbol{\beta}}_u(t)\;,$$

$$\frac{d\widetilde{\alpha}_u}{dt}(t)=-\mathbf{d}^0(t)^\intercal\widetilde{\boldsymbol{\beta}}_u(t)-\frac{1}{2}(\widetilde{\boldsymbol{\beta}}_u(t))^\intercal\mathbf{V}^0(t)\widetilde{\boldsymbol{\beta}}_u(t)\;,$$

with terminal conditions $\widetilde{\alpha}_u(u)=0$ and $\widetilde{\boldsymbol{\beta}}_u(u)=0$.

The functions $\widetilde{\alpha}_u^{k,l}$, $\widetilde{\boldsymbol{\beta}}_u^{k,l}$, $\widehat{\alpha}_u^{k,l}$, and $\widehat{\boldsymbol{\beta}}_u^{k,l}$, $k, l \in \mathcal{K}$, corresponding to $\widetilde{\alpha}_u$ and $\widetilde{\boldsymbol{\beta}}_u$ or to $\alpha_u^{i,j}$ and $\boldsymbol{\beta}_u^{i,j}$ for $i, j \in \mathcal{K}$ as considered in equations (4.29), (4.30), (4.31) and (4.32), then solve the ODEs

$$\frac{d\widetilde{\boldsymbol{\beta}}_u^{k,l}}{dt}(t) = -\mathbf{d}^1(t)^\mathsf{T}\widetilde{\boldsymbol{\beta}}_u^{k,l}(t) - (\widetilde{\boldsymbol{\beta}}_u(t))^\mathsf{T}\mathbf{V}^1(t)\widetilde{\boldsymbol{\beta}}_u^{k,l}(t),$$

$$\frac{d\widetilde{\alpha}_u^{k,l}}{dt}(t) = -\mathbf{d}^0(t)^\mathsf{T}\widetilde{\boldsymbol{\beta}}_u^{k,l}(t) - (\widetilde{\boldsymbol{\beta}}_u(t))^\mathsf{T}\mathbf{V}^0(t)\widetilde{\boldsymbol{\beta}}_u^{k,l}(t),$$

with terminal conditions $\widetilde{\alpha}_u^{k,l}(u) = c^{k,l}$, $\widetilde{\boldsymbol{\beta}}_u^{k,l}(u) = \mathbf{b}^{k,l}$ and

$$\frac{d\widehat{\boldsymbol{\beta}}_u^{k,l}}{dt}(t) = -\mathbf{d}^1(t)^\mathsf{T}\widehat{\boldsymbol{\beta}}_u^{k,l}(t) - (\boldsymbol{\beta}_u^{i,j}(t))^\mathsf{T}\mathbf{V}^1(t)\widehat{\boldsymbol{\beta}}_u^{k,l}(t),$$

$$\frac{d\widehat{\alpha}_u^{k,l}}{dt}(t) = -\mathbf{d}^0(t)^\mathsf{T}\widehat{\boldsymbol{\beta}}_u^{k,l}(t) - (\boldsymbol{\beta}_u^{i,j}(t))^\mathsf{T}\mathbf{V}^0(t)\widehat{\boldsymbol{\beta}}_u^{k,l}(t),$$

with terminal conditions $\widehat{\alpha}_u^{k,l}(u) = c^{k,l}$, $\widehat{\boldsymbol{\beta}}_u^{k,l}(u) = \mathbf{b}^{k,l}$.

Note that with these specifications, we obtain by (4.24) that for every $0 \leq t < u \leq T$ and every $i, j \in \mathcal{K}$, $i \neq j$

$$\mathbb{E}\left[p_{i,j}(0, u) \mid \mathcal{F}_t\right] = 1 - e^{-\int_0^t \psi_v^{i,j} dv}\mathbb{E}\left[e^{-\int_t^u \psi_v^{i,j} dv} \;\middle|\; \mathcal{F}_t\right] = 1 - e^{-\int_0^t \psi_v^{i,j} dv}e^{\alpha_u^{i,j}(t) + (\boldsymbol{\beta}_u^{i,j}(t))^\mathsf{T}\boldsymbol{\mu}_t}$$

$$= 1 - e^{\alpha_u^{i,j}(0) + (\boldsymbol{\beta}_u^{i,j}(0))^\mathsf{T}\boldsymbol{\mu}_0} - \int_0^t e^{-\int_0^s \psi_v^{i,j} dv}e^{\alpha_u^{i,j}(s) + (\boldsymbol{\beta}_u^{i,j}(s))^\mathsf{T}\boldsymbol{\mu}_s}(\boldsymbol{\sigma}(s, \boldsymbol{\mu}_s))^\mathsf{T}\boldsymbol{\beta}_u^{i,j}(s)d\mathbf{W}_s$$

and

$$\mathbb{E}\left[p_{i,i}(0, u) \mid \mathcal{F}_t\right] = 1 - \sum_{\substack{j=1 \\ j \neq i}}^N \mathbb{E}\left[1 - e^{-\int_t^u \psi_v^{i,j} dv} \;\middle|\; \mathcal{F}_t\right] = 2 - N + \sum_{\substack{j=1 \\ j \neq i}}^N e^{-\int_0^t \psi_v^{i,j} dv}e^{\alpha_u^{i,j}(t) + (\boldsymbol{\beta}_u^{i,j}(t))^\mathsf{T}\boldsymbol{\mu}_t}$$

$$= 2 - N + \sum_{\substack{j=1 \\ j \neq i}}^N \left(e^{\alpha_u^{i,j}(0) + (\boldsymbol{\beta}_u^{i,j}(0))^\mathsf{T}\boldsymbol{\mu}_0} + \int_0^t e^{-\int_0^s \psi_v^{i,j} dv}e^{\alpha_u^{i,j}(s) + (\boldsymbol{\beta}_u^{i,j}(s))^\mathsf{T}\boldsymbol{\mu}_s}(\boldsymbol{\sigma}(s, \boldsymbol{\mu}_s))^\mathsf{T}\boldsymbol{\beta}_u^{i,j}(s)d\mathbf{W}_s\right).$$

Moreover, as every martingale with respect to the Brownian filtration $\mathbb{F}^W$ is continuous, we have for $0 \leq u \leq t \leq T$ that

$$\mathbb{E}\left[p_{i,j}(0, u) \mid \mathcal{F}_t\right] = p_{i,j}(0, u) = \lim_{w \nearrow u} \mathbb{E}\left[p_{i,j}(0, u) \mid \mathcal{F}_w\right]$$

$$= \lim_{w \nearrow u}\left(1 - e^{\alpha_u^{i,j}(0) + (\boldsymbol{\beta}_u^{i,j}(0))^\mathsf{T}\boldsymbol{\mu}_0} - \int_0^w e^{-\int_0^s \psi_v^{i,j} dv}e^{\alpha_u^{i,j}(s) + (\boldsymbol{\beta}_u^{i,j}(s))^\mathsf{T}\boldsymbol{\mu}_s}(\boldsymbol{\sigma}(s, \boldsymbol{\mu}_s))^\mathsf{T}\boldsymbol{\beta}_u^{i,j}(s)d\mathbf{W}_s\right)$$

$$= 1 - e^{\alpha_u^{i,j}(0) + (\boldsymbol{\beta}_u^{i,j}(0))^\mathsf{T}\boldsymbol{\mu}_0} - \int_0^u e^{-\int_0^s \psi_v^{i,j} dv}e^{\alpha_u^{i,j}(s) + (\boldsymbol{\beta}_u^{i,j}(s))^\mathsf{T}\boldsymbol{\mu}_s}(\boldsymbol{\sigma}(s, \boldsymbol{\mu}_s))^\mathsf{T}\boldsymbol{\beta}_u^{i,j}(s)d\mathbf{W}_s,$$

and

$$\mathbb{E}\left[p_{i,i}(0,u)\,|\,\mathcal{F}_t\right] = p_{i,i}(0,u) = \lim_{w\nearrow u}\mathbb{E}\left[p_{i,i}(0,u)\,|\,\mathcal{F}_w\right]$$

$$=2-N+\sum_{\substack{j=1\\j\neq i}}^{N}\left(e^{\alpha_u^{i,j}(0)+(\boldsymbol{\beta}_u^{i,j}(0))^{\mathsf{T}}\boldsymbol{\mu}_0}+\int_0^u e^{-\int_0^s\psi_v^{i,j}dv}e^{\alpha_u^{i,j}(s)+(\boldsymbol{\beta}_u^{i,j}(s))^{\mathsf{T}}\boldsymbol{\mu}_s}(\boldsymbol{\sigma}(s,\boldsymbol{\mu}_s))^{\mathsf{T}}\boldsymbol{\beta}_u^{i,j}(s)d\mathbf{W}_s\right).$$

Hence, for arbitrary $u,t\in[0,T]$, we obtain

$$\mathbb{E}\left[p_{i,j}(0,u)\,|\,\mathcal{F}_t\right] = c_1^{i,j}(u) + \int_0^t \vartheta_1^{i,j}(s,u)\mathbb{1}_{[0,u]}(s)d\mathbf{W}_s\,, \tag{4.33}$$

$$\mathbb{E}\left[p_{i,i}(0,u)\,|\,\mathcal{F}_t\right] = c_2^i(u) + \int_0^t \vartheta_2^i(s,u)\mathbb{1}_{[0,u]}(s)d\mathbf{W}_s\,, \tag{4.34}$$

where for $u,s\in[0,T]$ and every $i,j\in\mathcal{K}$, $i\neq j$

$$c_1^{i,j}(u) := 1 - e^{\alpha_u^{i,j}(0)+(\boldsymbol{\beta}_u^{i,j}(0))^{\mathsf{T}}\boldsymbol{\mu}_0}\,,$$

$$\vartheta_1^{i,j}(s,u) := -e^{-\int_0^s\psi_v^{i,j}dv}e^{\alpha_u^{i,j}(s)+(\boldsymbol{\beta}_u^{i,j}(s))^{\mathsf{T}}\boldsymbol{\mu}_s}(\boldsymbol{\sigma}(s,\boldsymbol{\mu}_s))^{\mathsf{T}}\boldsymbol{\beta}_u^{i,j}(s)\,,$$

$$c_2^i(u) := 2 - N + \sum_{\substack{j=1\\j\neq i}}^{N}e^{\alpha_u^{i,j}(0)+(\boldsymbol{\beta}_u^{i,j}(0))^{\mathsf{T}}\boldsymbol{\mu}_0}\,,$$

$$\vartheta_2^i(s,u) := \sum_{\substack{j=1\\j\neq i}}^{N}e^{-\int_0^s\psi_v^{i,j}dv}e^{\alpha_u^{i,j}(s)+(\boldsymbol{\beta}_u^{i,j}(s))^{\mathsf{T}}\boldsymbol{\mu}_s}(\boldsymbol{\sigma}(s,\boldsymbol{\mu}_s))^{\mathsf{T}}\boldsymbol{\beta}_u^{i,j}(s)\,.$$

Similarly we get for $0\leq t<u\leq T$ and every $i,j,k\in\mathcal{K}$ with $i\neq j$, $j\neq k$ that

$$\mathbb{E}\left[p_{i,j}(0,u)\psi_u^{j,k}\,|\,\mathcal{F}_t\right] = \mathbb{E}\left[\psi_u^{j,k}\,|\,\mathcal{F}_t\right] - e^{-\int_0^t\psi_v^{i,j}dv}\mathbb{E}\left[e^{-\int_t^u\psi_v^{i,j}dv}\psi_u^{j,k}\,\Big|\,\mathcal{F}_t\right]$$

$$=e^{\widetilde{\alpha}_u(t)+(\widetilde{\boldsymbol{\beta}}_u(t))^{\mathsf{T}}\boldsymbol{\mu}_t}(\widetilde{\alpha}_u^{j,k}(t)+(\widetilde{\boldsymbol{\beta}}_u^{j,k}(t))^{\mathsf{T}}\boldsymbol{\mu}_t) - e^{-\int_0^t\psi_v^{i,j}dv}e^{\alpha_u^{i,j}(t)+(\boldsymbol{\beta}_u^{i,j}(t))^{\mathsf{T}}\boldsymbol{\mu}_t}(\widehat{\alpha}_u^{j,k}(t)+(\widehat{\boldsymbol{\beta}}_u^{j,k}(t))^{\mathsf{T}}\boldsymbol{\mu}_t)$$

$$=e^{\widetilde{\alpha}_u(0)+(\widetilde{\boldsymbol{\beta}}_u(0))^{\mathsf{T}}\boldsymbol{\mu}_0}(\widetilde{\alpha}_u^{j,k}(0)+(\widetilde{\boldsymbol{\beta}}_u^{j,k}(0))^{\mathsf{T}}\boldsymbol{\mu}_0) - e^{\alpha_u^{i,j}(0)+(\boldsymbol{\beta}_u^{i,j}(0))^{\mathsf{T}}\boldsymbol{\mu}_0}(\widehat{\alpha}_u^{j,k}(0)+(\widehat{\boldsymbol{\beta}}_u^{j,k}(0))^{\mathsf{T}}\boldsymbol{\mu}_0')$$

$$+\int_0^t\Big\{e^{\widetilde{\alpha}_u(s)+(\widetilde{\boldsymbol{\beta}}_u(s))^{\mathsf{T}}\boldsymbol{\mu}_s}(\boldsymbol{\sigma}(s,\boldsymbol{\mu}_s))^{\mathsf{T}}\big((\widetilde{\alpha}_u^{j,k}(s)+(\widetilde{\boldsymbol{\beta}}_u^{j,k}(s))^{\mathsf{T}}\boldsymbol{\mu}_s)\widetilde{\boldsymbol{\beta}}_u(s)+\widetilde{\boldsymbol{\beta}}_u^{j,k}(s)\big)$$

$$+e^{-\int_0^s\psi_v^{j,k}dv}e^{\alpha_u^{i,j}(s)+(\boldsymbol{\beta}_u^{i,j}(s))^{\mathsf{T}}\boldsymbol{\mu}_s}(\widehat{\alpha}_u^{j,k}(s)+(\widehat{\boldsymbol{\beta}}_u^{j,k}(s))^{\mathsf{T}}\boldsymbol{\mu}_s)(\boldsymbol{\sigma}(s,\boldsymbol{\mu}_s))^{\mathsf{T}}\boldsymbol{\beta}_u^{i,j}(s)$$

$$+e^{-\int_0^s\psi_v^{j,k}dv}e^{\alpha_u^{i,j}(s)+(\boldsymbol{\beta}_u^{i,j}(s))^{\mathsf{T}}\boldsymbol{\mu}_s}(\boldsymbol{\sigma}(s,\boldsymbol{\mu}_s))^{\mathsf{T}}\widehat{\boldsymbol{\beta}}_u^{j,k}(s)\Big\}d\mathbf{W}_s$$

and

$$
\mathbb{E}\left[p_{j,j}(0,u)\psi_u^{j,k}\,\middle|\,\mathcal{F}_t\right] = (2-N)\mathbb{E}\left[\psi_u^{j,k}\,\middle|\,\mathcal{F}_t\right] + \sum_{\substack{l=1\\l\neq j}}^{N} e^{-\int_0^t \psi_v^{j,l}dv}\mathbb{E}\left[e^{-\int_t^u \psi_v^{j,l}dv}\psi_u^{j,k}\,\middle|\,\mathcal{F}_t\right]
$$

$$
=(2-N)e^{\widetilde{\alpha}_u(0)+(\widetilde{\beta}_u(0))^\intercal\boldsymbol{\mu}_0}(\widetilde{\alpha}_u^{j,k}(0) + (\widehat{\boldsymbol{\beta}}_u^{j,k}(0))^\intercal\boldsymbol{\mu}_0) + \sum_{\substack{l=1\\l\neq j}}^{N} e^{\alpha_u^{j,l}(0)+(\beta_u^{j,l}(0))^\intercal\boldsymbol{\mu}_0}(\widehat{\alpha}_u^{j,k}(0) + (\widehat{\boldsymbol{\beta}}_u^{j,k}(0))^\intercal\boldsymbol{\mu}_0)
$$

$$
+ \int_0^t (2-N)\Big\{e^{\widetilde{\alpha}_u(s)+(\widetilde{\beta}_u(s))^\intercal\boldsymbol{\mu}_s}(\boldsymbol{\sigma}(s,\boldsymbol{\mu}_s))^\intercal\big((\widetilde{\alpha}_u^{j,k}(s) + (\widehat{\boldsymbol{\beta}}_u^{j,k}(s))^\intercal\boldsymbol{\mu}_s)\widetilde{\boldsymbol{\beta}}_u(s) + \widetilde{\boldsymbol{\beta}}_u^{j,k}(s)\big)
$$

$$
+ \sum_{\substack{l=1\\l\neq j}}^{N} e^{-\int_0^s \psi_v^{j,l}dv}e^{\alpha_u^{j,l}(s)+(\beta_u^{j,l}(s))^\intercal\boldsymbol{\mu}_s}(\widehat{\alpha}_u^{j,k}(s) + (\widehat{\boldsymbol{\beta}}_u^{j,k}(s))^\intercal\boldsymbol{\mu}_s)(\boldsymbol{\sigma}(s,\boldsymbol{\mu}_s))^\intercal\boldsymbol{\beta}_u^{j,l}(s)
$$

$$
+ e^{-\int_0^s \psi_v^{j,l}dv}e^{\alpha_u^{j,l}(s)+(\beta_u^{j,l}(s))^\intercal\boldsymbol{\mu}_s}(\boldsymbol{\sigma}(s,\boldsymbol{\mu}_s))^\intercal\widehat{\boldsymbol{\beta}}_u^{j,k}(s)\Big\}d\mathbf{W}_s.
$$

Note that by Jensen's inequality and Assumption 4.4.1 4), we get for every $0 \leq t < u \leq T$ and every $i,j,k \in \mathcal{K}$ with $j \neq k$ that

$$
\mathbb{E}\left[\mathbb{E}\left[p_{i,j}(0,u)\psi_u^{j,k}\,\middle|\,\mathcal{F}_t\right]^2\right] \leq \mathbb{E}\left[\mathbb{E}\left[p_{i,j}(0,u)^2(\psi_u^{j,k})^2\,\middle|\,\mathcal{F}_t\right]\right] \leq \mathbb{E}\left[(\psi_u^{j,k})^2\right] \leq C.
$$

Finally, with the same limit-arguments as above, we obtain for arbitrary $u,t \in [0,T]$ and every $i,j,k \in \mathcal{K}$ with $i \neq j$, $j \neq k$ that

$$
\mathbb{E}\left[p_{i,j}(0,u)\psi_u^{j,k}\,\middle|\,\mathcal{F}_t\right] = c_3(u) + \int_0^t \vartheta_3^{i,j,k}(s,u)\mathbb{1}_{[0,u]}(s)d\mathbf{W}_s, \tag{4.35}
$$

$$
\mathbb{E}\left[p_{j,j}(0,u)\psi_u^{j,k}\,\middle|\,\mathcal{F}_t\right] = c_4(u) + \int_0^t \vartheta_4^{j,k}(s,u)\mathbb{1}_{[0,u]}(s)d\mathbf{W}_s, \tag{4.36}
$$

where for $u,s \in [0,T]$, $i,j,k \in \mathcal{K}$ with $i \neq j$, $j \neq k$

$$
c_3(u) := e^{\widetilde{\alpha}_u(0)+(\widetilde{\beta}_u(0))^\intercal\boldsymbol{\mu}_0}(\widetilde{\alpha}_u^{j,k}(0) + (\widehat{\boldsymbol{\beta}}_u^{j,k}(0))^\intercal\boldsymbol{\mu}_0) - e^{\alpha_u^{i,j}(0)+(\beta_u^{i,j}(0))^\intercal\boldsymbol{\mu}_0}(\widehat{\alpha}_u^{j,k}(0) + (\widehat{\boldsymbol{\beta}}_u^{j,k}(0))^\intercal\boldsymbol{\mu}_0),
$$

$$
\vartheta_3^{i,j,k}(u,s) := \Big\{e^{\widetilde{\alpha}_u(s)+(\widetilde{\beta}_u(s))^\intercal\boldsymbol{\mu}_s}(\boldsymbol{\sigma}(s,\boldsymbol{\mu}_s))^\intercal\big((\widetilde{\alpha}_u^{j,k}(s) + (\widehat{\boldsymbol{\beta}}_u^{j,k}(s))^\intercal\boldsymbol{\mu}_s)\widetilde{\boldsymbol{\beta}}_u(s) + \widetilde{\boldsymbol{\beta}}_u^{j,k}(s)\big)
$$

$$
+ e^{-\int_0^s \psi_v^{j,k}dv}e^{\alpha_u^{i,j}(s)+(\beta_u^{i,j}(s))^\intercal\boldsymbol{\mu}_s}(\boldsymbol{\sigma}(s,\boldsymbol{\mu}_s))^\intercal\big((\widehat{\alpha}_u^{j,k}(s) + (\widehat{\boldsymbol{\beta}}_u^{j,k}(s))^\intercal\boldsymbol{\mu}_s)\beta_u^{i,j}(s) + \widehat{\boldsymbol{\beta}}_u^{j,k}(s)\big)\Big\},
$$

$$
c_4(u) := (2-N)e^{\widetilde{\alpha}_u(0)+(\widetilde{\beta}_u(0))^\intercal\boldsymbol{\mu}_0}(\widetilde{\alpha}_u^{j,k}(0) + (\widehat{\boldsymbol{\beta}}_u^{j,k}(0))^\intercal\boldsymbol{\mu}_0)
$$

$$
+ \sum_{\substack{l=1\\l\neq j}}^{N} e^{\alpha_u^{j,l}(0)+(\beta_u^{j,l}(0))^\intercal\boldsymbol{\mu}_0}(\widehat{\alpha}_u^{j,k}(0) + (\widehat{\boldsymbol{\beta}}_u^{j,k}(0))^\intercal\boldsymbol{\mu}_0),
$$

$$\vartheta_4^{j,k}(u,s) := \left\{ (2-N)e^{\widetilde{\alpha}_u(s)+(\widetilde{\beta}_u(s))^\mathsf{T}\boldsymbol{\mu}_s}(\boldsymbol{\sigma}(s,\boldsymbol{\mu}_s))^\mathsf{T}\left((\widetilde{\alpha}_u^{j,k}(s)+(\widehat{\boldsymbol{\beta}}_u^{j,k}(s))^\mathsf{T}\boldsymbol{\mu}_s)\widetilde{\beta}_u(s)+\widetilde{\boldsymbol{\beta}}_u^{j,k}(s)\right) \right.$$

$$\left. + \sum_{\substack{l=1 \\ l\neq j}}^{N} e^{-\int_0^s \psi_v^{j,l}dv}e^{\alpha_u^{j,l}(s)+(\beta_u^{j,l}(s))^\mathsf{T}\boldsymbol{\mu}_s}(\boldsymbol{\sigma}(s,\boldsymbol{\mu}_s))^\mathsf{T}\left((\widehat{\alpha}_u^{j,k}(s)+(\widehat{\boldsymbol{\beta}}_u^{j,k}(s))^\mathsf{T}\boldsymbol{\mu}_s)\beta_u^{j,l}(s)+\widehat{\boldsymbol{\beta}}_u^{j,k}(s)\right) \right\}.$$

We can now apply these results to compute explicitly the risk-minimizing hedging strategy as given in Theorem 4.3.3. From Equations (4.33), (4.34), (4.35) and (4.36) it follows immediately that the processes $\xi^{i,j}(u,\cdot)$ and $\theta^{i,j,k}(u,\cdot)$, $i,j,k \in \mathcal{K}$, $j \neq k$, of (4.16) and (4.17) are given as

$$\xi^{i,j}(u,t) = \vartheta_1^{i,j}(u,t), \quad \xi^{i,i}(u,t) = \vartheta_2^{i}(u,t),$$

$$\theta^{i,j,k}(u,t) = \vartheta_3^{i,j,k}(u,t), \quad \theta^{j,j,k}(u,t) = \vartheta_4^{j,k}(u,t).$$

For notational convenience we omit the explicit mentioning of the resulting risk-minimizing strategies in the affine setting here.

Note that with Equations (4.29), (4.30), (4.31) and (4.32) the i-th component $F^i(t,T)$, $i \in \mathcal{K}$, of $\mathbf{F}(t,T)$ in (4.13) is given as

$$F^i(t,T) = \frac{S^*(t)}{S^*(T)}Y^i\mathbb{E}\left[p_{i,i}(t,T) \mid \mathcal{F}_t\right] + \sum_{\substack{j=1 \\ j\neq i}}^{N}\frac{S^*(t)}{S^*(T)}Y^j\mathbb{E}\left[p_{i,j}(t,T) \mid \mathcal{F}_t\right]$$

$$+ \int_t^T \frac{S^*(t)}{S^*(u)}\mathbb{E}\left[p_{i,i}(t,u) \mid \mathcal{F}_t\right]\nu^i(u)du + \sum_{\substack{j=1 \\ j\neq i}}^{N}\int_t^T \frac{S^*(t)}{S^*(u)}\mathbb{E}\left[p_{i,j}(t,u) \mid \mathcal{F}_t\right]\nu^j(u)du$$

$$+ \sum_{\substack{k=1 \\ k\neq i}}^{N}\int_t^T \frac{S^*(t)}{S^*(u)}Z^{i,k}(u)\mathbb{E}\left[p_{i,i}(t,u)\psi^{i,k}(u) \mid \mathcal{F}_t\right]du$$

$$+ \sum_{\substack{j=1 \\ j\neq i}}^{N}\sum_{\substack{k=1 \\ k\neq j}}^{N}\int_t^T \frac{S^*(t)}{S^*(u)}Z^{j,k}(u)\mathbb{E}\left[p_{i,j}(t,u)\psi^{j,k}(u) \mid \mathcal{F}_t\right]du$$

$$= \frac{S^*(t)}{S^*(T)}Y^i\left(2-N+\sum_{\substack{j=1 \\ j\neq i}}^{N}e^{\alpha^{i,j}(t,T)+(\beta^{i,j}(t,T))^\mathsf{T}\boldsymbol{\mu}_t}\right) + \sum_{\substack{j=1 \\ j\neq i}}^{N}\frac{S^*(t)}{S^*(T)}Y^j\left(1-e^{\alpha^{i,j}(t,T)+(\beta^{i,j}(t,T))^\mathsf{T}\boldsymbol{\mu}_t}\right)$$

$$+ \int_t^T \frac{S^*(t)}{S^*(u)}\left(2-N+\sum_{\substack{j=1 \\ j\neq i}}^{N}e^{\alpha^{i,j}(t,u)+(\beta^{i,j}(t,u))^\mathsf{T}\boldsymbol{\mu}_t}\right)\nu^i(u)du$$

$$+ \sum_{\substack{j=1 \\ j\neq i}}^{N}\int_t^T \frac{S^*(t)}{S^*(u)}\left(1-e^{\alpha_t^{i,j}(s)+(\beta_t^{i,j}(s))^\mathsf{T}\boldsymbol{\mu}_s}\right)\nu^j(u)du$$

$$+ \sum_{\substack{k=1 \\ k \neq i}}^{N} \int_t^T \frac{S^*(t)}{S^*(u)} Z^{i,k}(u) \Bigg\{ (2 - N) e^{\widetilde{\alpha}(t,u) + (\widetilde{\boldsymbol{\beta}}(t,u))^\mathsf{T} \boldsymbol{\mu}_t} \big(\widetilde{\alpha}^{j,k}(t,u) + (\widetilde{\boldsymbol{\beta}}^{i,k}(t,u))^\mathsf{T} \boldsymbol{\mu}_t \big)$$

$$+ \sum_{\substack{l=1 \\ l \neq j}}^{N} e^{\alpha^{i,l}(t,u) + (\boldsymbol{\beta}^{i,l}(t,u))^\mathsf{T} \boldsymbol{\mu}_t} \big(\widehat{\alpha}^{i,k}(t,u) + (\widehat{\boldsymbol{\beta}}^{i,k}(t,u))^\mathsf{T} \boldsymbol{\mu}_t \big) \Bigg\} du$$

$$+ \sum_{\substack{j=1 \\ j \neq i}}^{N} \sum_{\substack{k=1 \\ k \neq j}}^{N} \int_t^T \frac{S^*(t)}{S^*(u)} Z^{j,k}(u) \Bigg\{ e^{\widetilde{\alpha}(t,u) + (\widetilde{\boldsymbol{\beta}}(t,u))^\mathsf{T} \boldsymbol{\mu}_t} \big(\widetilde{\alpha}^{j,k}(t,u) + (\widetilde{\boldsymbol{\beta}}^{j,k}(t,u))^\mathsf{T} \boldsymbol{\mu}_t \big)$$

$$- e^{\alpha^{i,j}(t,u) + (\boldsymbol{\beta}^{i,j}(t,u))^\mathsf{T} \boldsymbol{\mu}_t} \big(\widehat{\alpha}^{j,k}(t,u) + (\widehat{\boldsymbol{\beta}}^{j,k}(t,u))^\mathsf{T} \boldsymbol{\mu}_t \big) \Bigg\} du \, ,$$

and can hence be expressed explicitly in terms of $\boldsymbol{\mu}$. In particular, we can determine the integrand $\boldsymbol{\beta}$, given in (4.12), explicitly in this setting.

5. Conclusion

5. Conclusion

In this thesis the creation of innovative mathematical models for premium determination and risk-mitigation has been addressed by elaborating suitable premium evaluation schemes for unemployment insurance products and optimal hedging strategies for a large set of currently adopted insurance policies. The major advantage of all frameworks is their very general underlying model assumptions allowing the specification of the results in different scenarios. This flexibility is obtained by adopting the class of $\mathbb{F}$-doubly stochastic Markov chains and their property of (usually) admitting matrix-valued stochastic intensity processes, adapted to some arbitrary reference filtration $\mathbb{F}$. This general intensity-based framework complements the well-known reduced-form or hazard-rate models by considering consecutive transitions between the states of a finite state-space.

All results are developed in top-down approaches, starting from sufficiently general model assumptions, which are then further specified step by step to provide specific results. This, of course, provides the interested reader to adopt partial results and investigate the respective problem adjusted to his own interests and information.

In the context of premium determination for unemployment insurance, the employment-unemployment progress of an insured person has been modeled as a two-state $\mathbb{F}$-doubly stochastic Markov chain with differently generated intensities. The investigation of the dependencies of the insurance premium within hybrid markets has, among others, intrinsically been provided by the use of the benchmark approaches real-world pricing formula. Here, the discounting factor is the $\mathbb{P}$-numéraire portfolio, a global indicator of market performance. Due to the optimality properties of this portfolio, particularly on the long run, the derived insurance premiums provide economically reasonable prices also for long-dated insurance contracts. The premium determination frameworks furthermore incorporate several covariate processes, allowing the insurance premium to be adjusted flexibly to individual-related as well as micro- and macro-economic risk-factors. The time-homogeneous specification of the $\mathbb{F}$-doubly stochastic Markov chain allows the insurance premium being specified in closed analytic expressions, depending on the paths of the $\mathbb{P}$-numéraire portfolio. The results in a general Lévy process setting for the $\mathbb{P}$-numéraire portfolio and a (classical) Markov chain setting show the large design flexibility of this framework. In the even more general setting of a time-inhomogeneous $\mathbb{F}$-doubly stochastic Markov chain, Cox's proportional hazards model has been applied to estimate the employment-unemployment intensities for the German labor market on a dataset provided by IAB, the German institute for employment research. In this context, the extension of the existing literature by showing the applicability of Cox's proportional hazards estimators to the intensities of $\mathbb{F}$-doubly stochastic Markov chains can be pointed out. The estimation results show the desired sufficiency for most of the standard goodness-of-fit methods as well as a proper predictive power according to a newly developed goodness-of-fit method. This allowed the calculation of the insurance premiums based on the estimated intensities by Monte Carlo simulations, providing an insurance premium calibrated to a large set of risk-factors on the German labor market.

In the context of risk mitigation, the Galtchouk-Kunita-Watanabe decompositions for a large set of insurance products have been elaborated in order to derive mean-variance and risk-minimizing hedging strategies with respect to different hedging instruments. A current topic of the life insurance and pension funds industry has been addressed in this context: the hedging of longevity risk with longevity bonds. Here, mean-variance optimal trading strategies have been derived for pure endowment, term insurance and general life annuities under specific, yet sufficiently general model assumptions. Although reduced-form models have already been applied frequently in the literature, the presented results with their specification to a general affine structure of the mortality intensity provides some valuable input to this topic. Moreover, the presented advantageous properties of the gratification annuity can justify its introduction on the life insurance market. A larger contribution to hedging of insurance products has been provided by deriving risk-minimizing hedging strategies for general insurance payment processes, covering most of the currently adopted forms of insurance. The top-down-derivation of both frameworks furthermore allowed addressing questions on systematic and unsystematic as well as hedgable and unhedgable parts of an insurance contract's risk.

Appendices

A. $\mathbb{F}$-Doubly Stochastic Markov Chains

The class of $\mathbb{F}$-doubly stochastic Markov chains has been introduced in Jakubowski and Niewęgłowski [61] and its appropriateness for pricing and hedging purposes was shown in Jakubowski and Niewęgłowski [62]. In this appendix we briefly overview the definitions relevant results which are used throughout Chapters 2, 3 and 4.

Let $X = (X_t)_{t \in [0,T]}$ be a right-continuous stochastic process on $(\Omega, \mathcal{G}, \mathbb{P})$ with finite state-space $\mathcal{K} := \{1, ..., N\}$. Let $\mathbb{F}^X$ be the augmented filtration generated by X and let $\mathbb{F}$ be some arbitrary reference filtration assumed to satisfy the usual conditions of right-continuity and augmentation by the $\mathbb{P}$-null sets with $\mathcal{F}_0 = \{\emptyset, \Omega\}$, see Protter [80]. We consider the filtration $\mathbb{G}$ to be the enlargement of $\mathbb{F}^X$ through $\mathbb{F}$, i.e. we assume $\mathbb{G} = \mathbb{F}^X \vee \mathbb{F}$. Note that also $\mathbb{G}$ satisfies the usual conditions and $\mathcal{G}_0 = \{\emptyset, \Omega\}$. We furthermore introduce the auxiliary filtration $\widetilde{\mathbb{G}} = (\widetilde{\mathcal{G}}_t)_{t \in [0,T]}$ with $\widetilde{\mathcal{G}}_t := \mathcal{F}_t^X \vee \mathcal{F}_T$.

Definition A.1. *The process* $X = (X_t)_{t \in [0,T]}$ *is called an* $\mathbb{F}$*-doubly stochastic Markov chain with state space* $\mathcal{K}$*, if there exists a family of stochastic matrices* $\boldsymbol{P} = \boldsymbol{P}(s,t) = [p_{j,k}(s,t)]_{j,k \in \mathcal{K}}$*,* $0 \le s \le t \le T$*, such that*

(1) the matrix $\boldsymbol{P}(s,t)$ *is* $\mathcal{F}_t$*-measurable, and* $\boldsymbol{P}(s,.)$ *is progressively measurable,*

(2) for every $j, k \in \mathcal{K}$ *we have*

$$\mathbb{1}_{\{X_s = j\}} \mathbb{P}(X_t = k \mid \widetilde{\mathcal{G}}_t) = \mathbb{1}_{\{X_s = j\}} p_{j,k}(s,t) . \tag{A.1}$$

The family $\boldsymbol{P}$ *is called the* conditional transition probability process *of* X*.*

Definition A.1 generalizes the notion of a continuous time Markov chain which itself is an $\mathbb{F}$-doubly stochastic Markov chain with $\mathcal{F}_t = \{\emptyset, \Omega\}$ for all $t \in [0,T]$. Other examples for $\mathbb{F}$-doubly stochastic Markov chains are compound Poisson processes or Cox processes, see Jakubowski and Niewęgłowski [61]. Note however that since the reference filtration $\mathbb{F}$ is not specified an $\mathbb{F}$-doubly stochastic Markov chain is not necessarily a Markov process according to the usual definition.

Definition A.2. *We say that a state* $k \in \mathcal{K}$ *absorbing, if* $p_{k,j}(s,t) = 0$ *for all* $0 \le s < t \le T$ *and all* $j \in \mathcal{K}$ *with* $j \ne k$.

The next proposition states that the conditional transition probability process $\boldsymbol{P}$ satisfies the well-known Chapman-Kolmogorov equations.

Proposition A.3. *Let* X *be an* $\mathbb{F}$*-doubly stochastic Markov chain with transition probability process* $\boldsymbol{P}$*, then for every* $0 \le s \le t \le u \le T$ *we have*

$$\boldsymbol{P}(s,u) = \boldsymbol{P}(s,t)\boldsymbol{P}(t,u) \quad a.s.. \tag{A.2}$$

Proof. See Jakubowski and Niewęgłowski [61, Theorem 3.6.]. □

Proposition A.4. *If X is an $\mathbb{F}$-doubly stochastic Markov chain, then for every bounded, $\mathcal{F}_T$-measurable random variable Y and for each $t \in [0, T]$ we have*

$$\mathbb{E}[Y \mid \mathcal{G}_t] = \mathbb{E}[Y \mid \mathcal{F}_t]. \tag{A.3}$$

Proof. See Jakubowski and Niewęgłowski [61, Proposition 3.4.].   □

Property (A.3) is well-known in the context of credit-risk and life-insurance analysis as Hypothesis (H) or the immersion property. Therefore, Proposition A.4 shows that Hypothesis (H) holds for every $\mathbb{F}$-doubly stochastic Markov chain.

As already stated in the introduction, an important property which makes the class of $\mathbb{F}$-doubly stochastic Markov chains interesting for applications is that they may admit matrix-valued stochastic intensity processes in the following sense.

Definition A.5. *An $\mathbb{F}$-doubly stochastic Markov chain X with state space $\mathcal{K}$ is said to have an* intensity, *if there exists an $\mathbb{F}$-adapted matrix-valued stochastic process $\boldsymbol{\Psi} = (\boldsymbol{\Psi}_t)_{t \in [0,T]}$ with $\boldsymbol{\Psi}_t = \left[\psi_t^{j,k}\right]_{j,k \in \mathcal{K}}$ such that*

1) $\boldsymbol{\Psi}$ is integrable, i.e. we have

$$\int_{0+}^{T} \sum_{j \in \mathcal{K}} |\psi_s^{j,j}| ds < \infty . \tag{A.4}$$

2) $\boldsymbol{\Psi}$ satisfies the following conditions:

$$\psi_t^{j,k} \geq 0 \quad \forall j, k \in \mathcal{K}, j \neq k, \quad \psi_t^{j,j} = -\sum_{k \neq j} \psi_t^{j,k} \quad \forall j \in \mathcal{K}, \ t \in [0, T], \tag{A.5}$$

$$\boldsymbol{P}(v, t) - \boldsymbol{I}_N = \int_{v+}^{t} \boldsymbol{P}(v, u) \boldsymbol{\Psi}(u) du \quad \forall v \leq t \quad (Kolmogorov \ forward \ equation). \tag{A.6}$$

$$\boldsymbol{P}(v, t) - \boldsymbol{I}_N = \int_{v+}^{t} \boldsymbol{\Psi}(u) \boldsymbol{P}(u, t) du \quad \forall v \leq t \quad (Kolmogorov \ backward \ equation),$$

A process $\boldsymbol{\Psi}$ satisfying the above conditions is called an intensity *of the $\mathbb{F}$-doubly stochastic Markov chain X.*

Corollary A.6. *If an $\mathbb{F}$-doubly stochastic Markov chain X has intensity, then the conditional transition probability process $\boldsymbol{P}$ is jointly continuous at (s, t) for $0 \leq s \leq t \leq T$.*

Proof. See Jakubowski and Niewęgłowski [61, Corollary 3.9.].   □

Theorem A.7. *Let $\widetilde{\boldsymbol{\Psi}} = (\widetilde{\boldsymbol{\Psi}}_t)_{t \in [0,T]}$ be an $\mathbb{F}$-adapted $N \times N$ matrix-valued stochastic process satisfying the conditions (A.4) and (A.5) of Definition A.18. Then there exists an $\mathbb{F}$-doubly stochastic Markov chain X with intensity $\widetilde{\boldsymbol{\Psi}}$.*

Proof. See Jakubowski and Niewęgłowski [61, Theorem 4.8.].   □

Theorem A.7 is the core result for our applications in Chapters 2, 3, and 4. It states that for every sufficiently structured $N \times N$-matrix-valued stochastic process $\widetilde{\Psi}$ there exists an F-doubly stochastic Markov chain X having intensity $\widetilde{\Psi}$. Moreover, the proof of the theorem intrinsically provides the simulation scheme for the paths of an F-doubly stochastic Markov chain corresponding to $\widetilde{\Psi}$. We use this result for the Monte Carlo simulations in Section 2.3.6.

The proof of Theorem A.7 relies on a martingale characterization which is given in the next theorem. Before we state this result, we first introduce some processes related to the F-doubly stochastic Markov chain X.

For $j \in \mathcal{K}$, let

$$H_t^j := \mathbb{1}_{\{X_t = j\}}, \quad t \in [0, T], \tag{A.7}$$

be the indicator function for X of being in state j at time t and denote by $\mathbf{H} = (\mathbf{H}_t)_{t \in [0,T]}$ with $\mathbf{H}_t = (H_t^1, ..., H_t^N)^\intercal$ the corresponding N-variate vector-valued process. Moreover, for $j, k \in \mathcal{K}$, $j \neq k$, let $N^{jk} = (N_t^{jk})_{t \in [0,T]}$ with

$$N_t^{jk} := \int_{0+}^{t} H_{u-}^j dH_u^k = \sum_{0 < u \leq t} H_{u-}^j \bigtriangleup H_u^k \tag{A.8}$$

define the counting processes of the jumps of X from state j to k up to time t.

Theorem A.8. *Let $X = (X_t)_{t \in [0,T]}$ be a stochastic process with state space $\mathcal{K}$ and $\Psi = (\Psi_t)_{t \in [0,T]}$ be a matrix-valued process, satisfying (A.4) and (A.5) of Definition A.18. The following conditions are equivalent:*

i) X is an F-doubly stochastic Markov chain with an intensity Ψ.

ii) The process $\boldsymbol{M} = (\boldsymbol{M}_t)_{t \in [0,T]}$ with

$$\boldsymbol{M}_t := \boldsymbol{H}_t - \int_{0+}^{t} \Psi_u^\intercal \boldsymbol{H}_u du, \tag{A.9}$$

is a $\widetilde{\mathbb{G}}$-local martingale.

iii) For $j, k \in \mathcal{K}$, $j \neq k$, the processes $M^{jk} = (M_t^{jk})_{t \in [0,T]}$ with

$$M_t^{jk} := N_t^{jk} - \int_{0+}^{t} H_u^j \psi_u^{j,k} du \tag{A.10}$$

are $\widetilde{\mathbb{G}}$-local martingales.

iv) The process $\boldsymbol{L} = (\boldsymbol{L}_t)_{t \in [0,T]}$ with

$$\boldsymbol{L}_t := \boldsymbol{Q}(0,t)^{\mathsf{T}} \boldsymbol{H}_t \; ,$$

is a $\widetilde{\mathbb{G}}$-local martingale. Here $\boldsymbol{Q}(0,t)$ is a unique solution to the random integral equation

$$d\boldsymbol{Q}(0,t) = -\boldsymbol{\Psi}_t \boldsymbol{Q}(0,t) dt, \;\; \boldsymbol{Q}(0,0) = \boldsymbol{I}_N,$$

Note that then

$$\boldsymbol{L}_t = \boldsymbol{H}_0 + \int_{0+}^{t} \boldsymbol{Q}^{\mathsf{T}}(0,u) d\boldsymbol{M}_u \tag{A.11}$$

Proof. See Jakubowski and Niewęgłowski [61, Theorem 4.1.]. $\square$

The equivalence in Theorem A.8 between statement (i) and statement (iii) provides the key for the connection of F-doubly stochastic Markov chains and the counting process theory underlying the estimation schemes of Cox's proportional hazards model which we provide in Section 2.3.1.

Remark A.9.   *1) An equivalent formulation of Theorem A.8 statement (ii) is that for every $i \in \mathcal{K}$ the process $M^i = (M_t^i)_{t \in [0,T]}$ with*

$$M_t^i = H_t^i - \int_{0+}^{t} \sum_{j=1}^{N} \Psi_u^{j,i} H_u^j du = H_t^i - \int_{0+}^{t} \sum_{j=1}^{N} \Psi_u^{X_u,i} du$$

is a $\widetilde{\mathbb{G}}$-local martingale.

2) For every $t \in [0,T]$, the matrix $\boldsymbol{Q}(0,t)$ is the unique inverse matrix of $\boldsymbol{P}(0,t)$. More generally, for $0 \leq s \leq t$, we denote by $\boldsymbol{Q}(s,t)$ the unique inverse matrix of $\boldsymbol{P}(s,t)$. The existence and further properties of the family $\boldsymbol{Q}(s,t)$, $0 \leq s \leq t$, is given in Jakubowski and Niewęgłowski [61].

It follows immediately from (A.2) that for every $0 \leq s \leq t \leq u \leq T$ we have

$$\boldsymbol{P}(t,u) = \boldsymbol{Q}(s,t)\boldsymbol{P}(s,u) \tag{A.12}$$

3) As the processes $\boldsymbol{M}$, $\boldsymbol{L}$, and M^{jk}, $j,k \in \mathcal{K}$, $j \neq k$, are $\mathbb{G}$-adapted, they are also $\mathbb{G}$-local martingales.

Corollary A.10. *For every $j, k \in \mathcal{K}$, $j \neq k$, and for every $t \in [0, T]$ we have*

$$[M^{jk}]_t = N_t^{jk} \, ,$$

$$\langle M^{jk} \rangle_t = \int_{0+}^{t} H_u^j \psi^{j,k}(u) du \, . \tag{A.13}$$

Moreover, with Remark A.9 we have for $j \in \mathcal{K}, t \in [0, T]$, we have

$$[M^j]_t = \sum_{0 < s \leq t} (\Delta H_s^j)^2 = \sum_{\substack{k=1 \\ k \neq j}}^{N} (N_t^{kj} + N_t^{jk}) \, ,$$

$$\langle M^j \rangle_t = \sum_{\substack{k=1 \\ k \neq j}}^{N} \int_{0+}^{t} H_u^k \psi^{k,j}(u) du - \int_{0+}^{t} H_u^j \psi^{j,j}(u) du \, . \tag{A.14}$$

Proof. As N^{jk} is of finite variation and $\left(\int_{0+}^{t} H_u^j \psi_u^{j,k} du \right)_{t \in [0,T]}$ is continuous and of finite variation, it follows from the definition of M^{jk}, that

$$[M^{jk}]_t = [N^{jk}]_t = \sum_{0 < s \leq t} (\Delta N_s^{jk})^2 = N_t^{jk}.$$

Moreover, since $\left(\int_{0+}^{t} H_u^j \psi_u^{j,k} du \right)_{t \in [0,T]}$ is continuous and $\mathbb{G}$-adapted and therefore $\mathbb{G}$-predictable, it follows that

$$\langle M^{jk} \rangle_t = \int_{0+}^{t} H_u^j \psi_u^{j,k} du \, ,$$

as it is the compensator of $[M^{ij}] = N^{ij}$.

Next we observe that H^j is of finite variation and $\left(\int_{0+} t \sum_{k \in \mathcal{K}} H_u^k \psi_u^{k,j} du \right)_{t \in [0,T]}$ is continuous and of finite variation. It follows that

$$[M^j]_t = \sum_{0 < s \leq t} (\Delta H_s^j)^2 = \sum_{\substack{k=1 \\ k \neq j}}^{N} N_t^{kj} + N_t^{jk} \, ,$$

as $\sum_{0 < s \leq t} (\Delta H_s^j)^2$ counts the jumps of X into and out of the state j up to time t. As $\left(\int_{0+}^{t} H_u^j \psi_u^{j,k} du \right)_{t \in [0,T]}$ is the compensator of N^{jk} it follows that

$$\langle M^j \rangle_t = \sum_{\substack{k=1 \\ k \neq j}}^{N} \left(\int_{0+}^{t} H_u^k \psi_u^{k,j} du + \int_{0+}^{t} H_u^j \psi_u^{j,k} du \right)$$

$$= \sum_{\substack{k=1 \\ k \neq j}}^{N} \int_{0+}^{t} H_u^k \psi_u^{k,j} du - \int_{0+}^{t} H_u^j \psi_u^{j,j} du \,,$$

where the last equality is given by (A.5). This ends the proof. $\qquad\square$

An important property of F-doubly stochastic Markov chains with intensity is that their jump times avoid all F-stopping times.

Proposition A.11. *Let X be an F-doubly stochastic Markov chain with intensity and jump times $\tau_0 := 0$ and*

$$\tau_k := \inf\{\tau_{k-1} < t \leq T : X_t \neq X_{\tau_{k-1}}\}. \tag{A.15}$$

Then every jump time τ_k, $k \geq 1$, avoids F-stopping times, i.e. $\mathbb{P}(\tau_k = \varrho) = 0$ for every F-stopping time ϱ, provided that $\tau_k < \infty$ a.s..

Proof. See Jakubowski and Niewęgłowski [61, Proposition 6.1.]. $\qquad\square$

The following proposition is an important result for the GKW-decompositions of general insurance claims which we provide in Section 4.2.

Proposition A.12. *Let X be an F-doubly stochastic Markov chain with intensity. Then the local martingale $\boldsymbol{M}$, given in (A.9), is strongly orthogonal to every F-local martingale N in the (strong) sense that for each component M^i, $i \in \mathcal{K}$, the quadratic covariation $[M^i, N]$ vanishes.*

Proof. First note that M^i is a discontinuous finite variation local martingale. Its jump times $(\widetilde{\tau}_j)_{j \geq 0}$ with $\widetilde{\tau}_0 := 0$ and

$$\widetilde{\tau}_j := \inf\{t > \widetilde{\tau}_{j-1} | M_{t-}^i \neq M_t^i\} \,, \quad j \geq 1$$

is a subsequence of the jump times $(\tau_k)_{k \geq 0}$ of X as given by (A.15). As the jump times of the càdlàg local martingale N are F-stopping times, the processes M^i and N have almost surely no common jumps due to Proposition A.11.

This implies that for all $t \in [0, T]$ we have

$$[M^i, N]_t = M_0 N_0 + \sum_{0 < s \leq t} \Delta M_s^i \Delta N_s = 0.$$

The result then follows by integration by parts and the fact that stochastic integration preserves the local martingale property. $\qquad\square$

A useful result for the sojourn-times of an F-doubly stochastic Markov chain X is given in the next proposition.

Proposition A.13. *Let X be an $\mathbb{F}$-doubly stochastic Markov chain with intensity Ψ and let $(\tau_k)_{k \geq 0}$ be the sequence of jump-times as given in* (A.15). *Provided that $\tau_k < \infty$ a.s., we have for all $j \in \mathcal{K}$, $t \in \mathbb{R}_+$*

$$\mathbb{P}(\tau_k - \tau_{k-1} > t \mid \widetilde{\mathcal{G}}_{\tau_{k-1}}) = e^{-\int_{\tau_{k-1}}^{\tau_{k-1}+t} \psi_u^{X_{\tau_{k-1}}} du}, \tag{A.16}$$

where for all $k \in \mathcal{K}$, $s \in [0,T]$, we denote $\psi_s^k := \psi_s^{k,k}$.

Proof. See Jakubowski and Niewęgłowski [61, Corollary 5.2.]. $\qquad\square$

In the following subsection we show that the commonly known reduced-form or intensity based models, widely used e.g. in credit risk or life insurance modeling, constitute a particular example of $\mathbb{F}$-doubly stochastic Markov chains and that most of their well known formulas follow immediately by the above given results.

A.1. Reduced-Form Models

Setting up a reduced-form or hazard-rate model one usually introduces a non-negative random time τ on $(\Omega, \mathcal{G}, \mathbb{P})$ with $\mathbb{P}(\tau = 0) = 0$ and $\mathbb{P}(\tau > t) > 0$ for all $t \in [0,T]$. One then sets $H_t := \mathbb{1}_{\{\tau \leq t\}}$, $t \in [0,T]$, and the filtration $\mathbb{G}$ as $\mathbb{G} = \mathbb{F}^H \vee \mathbb{F}$ for some arbitrary reference filtration $\mathbb{F}$, satisfying the usual conditions and for $\mathbb{F}^H$ being the augmented filtration generated by $(H_t)_{t \in [0,T]}$.

We can translate this setting into the above mentioned framework by letting X be an $\mathbb{F}$-doubly stochastic Markov chain with state space $\{1, 2\}$. We assume the state 2 to be an absorbing state according to Definition A.2 and $\mathbb{P}(X_0 = 1) = 1$. Then τ is the first and only jump time τ_1 of X from state 1 to 2 according to (A.15). We then obviously have $H_t = H_t^2 = 1 - H_t^1$, $t \in [0,T]$, and $\mathbb{F}^X = \mathbb{F}^H$.

According to (A.1), we have for all $t \in [0,T]$ that

$$G_t := \mathbb{P}(\tau > t \mid \mathcal{F}_t) = \mathbb{E}\left[\mathbb{1}_{\{\tau > 0\}} \mathbb{P}(\tau > t \mid \widetilde{\mathcal{G}}_0) \mid \mathcal{F}_t\right] = p_{0,0}(0, t) \tag{A.17}$$

$$= \mathbb{E}\left[\mathbb{1}_{\{\tau > 0\}} \mathbb{P}(\tau > t \mid \widetilde{\mathcal{G}}_0) \mid \mathcal{F}_T\right] = \mathbb{P}(\tau > t \mid \mathcal{F}_T),$$

which confirms that Hypothesis (H), as indicated by Proposition A.4 indeed holds, see Bielecki and Rutkowski [24].

The process $G = (G_t)_{t \in [0,T]}$ is called the *conditional survival probability* process. If one assumes that $G_t > 0$ for all $t \in [0,T]$, then one can define the so called $\mathbb{F}$-hazard process $\Gamma = (\Gamma_t)_{t \in [0,T]}$ as

$$\Gamma_t := -\ln G_t. \tag{A.18}$$

In particular, we have $G_t = \mathbb{P}(\tau > t \mid \mathcal{F}_t) = e^{-\Gamma_t}$ and if one assumes that $\Gamma_t = \int_0^t \mu_s ds$ for some $\mathbb{F}$-progressively measurable process $\mu = (\mu_t)_{t \in [0,T]}$, then

$$G_t := \mathbb{P}(\tau > t \,|\, \mathcal{F}_t) = e^{-\int_0^t \mu_s ds}.$$

The process μ in this context is called the F-intensity of τ.

To connect this with the definition of an intensity for the F-doubly stochastic Markov chain X, we additionally assume X to have a 2×2-matrix valued intensity $\boldsymbol{\Psi}$. Because the state 2 is absorbing all of its entries of the second row need to equal 0. Due to this particular form of the intensity process, the solution to the Kolmogorov forward equation (A.6) can explicitly be obtained as a matrix valued stochastic exponential, see Dietz [46]. In particular, we get for all $0 \le s \le t$ that

$$\mathbf{P}(s,t) = \begin{pmatrix} e^{-\int_s^t \psi_u^{1,2} du} & 1 - e^{-\int_s^t \psi_u^{1,2} du} \\ 0 & 1 \end{pmatrix}.$$

Equation (A.17) then shows that $G_t = \mathbb{P}(\tau > t \,|\, \mathcal{F}_t) = e^{-\int_0^t \psi_u^{1,2} du}$. Therefore, the presence of an intensity $\boldsymbol{\Psi}$ corresponding to X guarantees that $G_t > 0$ and that $\psi^{1,2}$ is the F-intensity of τ.

As for every $0 \le s \le t$ the matrix $\mathbf{Q}(s,t)$ is the inverse matrix of $\mathbf{P}(s,t)$, we furthermore obtain that

$$\mathbf{Q}(s,t) = \begin{pmatrix} e^{\int_s^t \psi_u^{1,2} du} & 1 - e^{\int_s^t \psi_u^{1,2} du} \\ 0 & 1 \end{pmatrix}.$$

Theorem A.8 and Remark A.9 now state that the processes $\mathbf{M}$ and $\mathbf{L}$ defined by

$$\begin{pmatrix} M_t^1 \\ M_t^2 \end{pmatrix} = \begin{pmatrix} \mathbb{1}_{\{\tau > t\}} + \int_{0+}^t \psi_u^{1,2} \mathbb{1}_{\{\tau > u\}} du \\ \mathbb{1}_{\{\tau \le t\}} - \int_{0+}^t \psi_u^{1,2} \mathbb{1}_{\{\tau > u\}} du \end{pmatrix}$$

$$= \begin{pmatrix} 1 - \left(\mathbb{1}_{\{\tau \le t\}} - \int_{0+}^t \psi_u^{1,2} \mathbb{1}_{\{\tau > u\}} du \right) \\ \mathbb{1}_{\{\tau \le t\}} - \int_{0+}^t \psi_u^{1,2} \mathbb{1}_{\{\tau > u\}} du \end{pmatrix} =: \begin{pmatrix} 1 - M_t \\ M_t \end{pmatrix}$$

and

$$\begin{pmatrix} L_t^1 \\ L_t^2 \end{pmatrix} = \begin{pmatrix} e^{\int_0^t \psi_u^{1,2} du} \mathbb{1}_{\{\tau > t\}} \\ 1 - e^{\int_0^t \psi_u^{1,2} du} \mathbb{1}_{\{\tau > t\}} \end{pmatrix} =: \begin{pmatrix} L_t \\ 1 - L_t \end{pmatrix}$$

are G-local martingales. These are well known results in the context of reduced-form modeling. It can actually be shown that $L = L^1$ is a G-martingale here. Moreover, since in this setting $[M^1]_t = H_t^1 \le 1$ and $[M^2]_t = H_t^2 \le 1$, we have that $M = M^2$ is a square integrable G-martingale.

Because $dM_t^1 = (1 - H_{t-})dM_t^1 = -(1 - H_{t-})dM_t^2$, we have due to Equation (A.11) that

$$dL_t = dL_t^1 = e^{\int_0^t \psi_u^{1,2} du} dM_u^1 = -e^{\int_0^t \psi_u^{1,2} du} \mathbb{1}_{\{\tau > t-\}} dM_u^2 = -L_{t-} dM_t, \qquad (A.19)$$

such that L is the stochastic exponential with respect to $-M$.

These results can also be generalized to the cases where we do not assume given an intensity but only the hazard process Γ. Here, the processes $M = (M_t)_{t \in [0,T]}$ with

$$M_t = H_t - \Gamma_{t \wedge \tau} \qquad (A.20)$$

and $L = (L_t)_{t \in [0,T]}$ with

$$L_t = e^{\Gamma_t} \mathbb{1}_{\{\tau > t\}} \qquad (A.21)$$

are $\mathbb{G}$-martingales and $dL_t = -L_{t-} dM_t$ holds as well, see Bielecki and Rutkowski [24].

To end this subsection, we provide formulas on conditional expectations which we use in Section 3.2.

Corollary A.14. *For $0 \leq t < s \leq T$ let Y be an $\mathcal{F}_s$-measurable random variable. Then we have*

$$\mathbb{E}\left[\mathbb{1}_{\{\tau > s\}} Y \mid \mathcal{G}_t\right] = \mathbb{1}_{\{\tau > t\}} \mathbb{E}\left[e^{\Gamma_t - \Gamma_s} Y \mid \mathcal{F}_t\right]. \qquad (A.22)$$

Proof. See Bielecki and Rutkowski [24, Corollary 5.1.1.]. □

Corollary A.15. *Let $h : \mathbb{R}^+ \to \mathbb{R}$ be a bounded, continuous function. Then for any $0 \leq t < s \leq T$ we have*

$$\mathbb{E}\left[\mathbb{1}_{\{t < \tau \leq s\}} h(\tau) \mid \mathcal{G}_t\right] = \mathbb{1}_{\{\tau > t\}} \mathbb{E}\left[\int_t^s h(u) e^{\Gamma_t - \Gamma_u} d\Gamma_u \mid \mathcal{F}_t\right] \qquad (A.23)$$

Proof. See Bielecki and Rutkowski [24, Corollary 5.1.3.]. □

A.2. Time Homogeneous $\mathbb{F}$-Doubly Stochastic Markov Chains

For the first application of $\mathbb{F}$-doubly stochastic Markov chains to pricing unemployment insurance, we assume the conditional transition probability process $\mathbf{P}$ to be time-homogeneous in the sense defined below. Broadly speaking, we want for every $0 \leq s \leq t \leq T$ and every $h > 0$ such that $t + h \leq T$ to have

$$\mathbf{P}(s + h, t + h) = \mathbf{P}(s, t).$$

For classical Markov chains, this is exactly the definition of time-homogeneity. The consideration of $\mathbb{F}$-doubly stochastic Markov chains, however, allows making assumptions on the measurability of $\mathbf{P}$. In Biagini and Widenmann [14], we slightly adjusted the general Definition A.1 to a time-homogeneous version as follows.

Definition A.16. *A stochastic process $X = (X_t)_{t\in[0,T]}$ is called time-homogeneous F-doubly stochastic Markov chain with state space $\mathcal{K}$ if there exists a matrix-valued family $\boldsymbol{P} = \big(\boldsymbol{P}(t)\big)_{t\in[0,T]} = \big(\big[p_{i,j}(t)\big]_{j,k\in\mathcal{K}}\big)_{t\in[0,T]}$, such that for every $0 \leq s \leq t \leq T$ and every $j,k \in \mathcal{K}$*

(i) $\boldsymbol{P}(t-s)$ is a stochastic matrix, i.e. the sum of all row entries is one.

(ii) $\boldsymbol{P}(t-s)$ is $\mathcal{F}_T$-measurable.

(iii) $\mathbb{1}_{\{X_s=j\}}\mathbb{P}\left(X_t = k\,\big|\,\widetilde{\mathcal{G}}_s\right) = \mathbb{1}_{\{X_s=i-1\}}p_{i,j}(t-s).$

Remark A.17. *1) There are two basic differences between Definition A.16 and Definition A.1. While in Definition A.1 we assume $\mathcal{F}_t$-measurability of the stochastic matrix $\boldsymbol{P}(s,t)$, we merely assume $\mathcal{F}_T$-measurability of $\boldsymbol{P}(t)$ in Definition A.16. This allows considering $\mathcal{F}_T$-measurable intensity matrices which do not vary in time, whereas the more general definition requests F-adapted matrix-valued intensity processes.*

The proceedings of the proofs as given in Jakubowski and Niewęgłowski [61] are despite the different definitions also applicable to the time-homogeneous case.

2) Due to the definition of an F-doubly stochastic Markov chain and Jakubowski and Niewęgłowski [61, Lemma 3.1], we have that the σ-fields $\mathcal{Z}_t^X := \sigma(X_u : u > t)$, $t \in [0,T]$, and $\mathcal{V}_t^X := \sigma(X_u : u < t)$, $t \in [0,T]$, are conditionally independent given $\mathcal{F}_T \vee \sigma(X_t)$.

3) Hypothesis (H) is valid also for the time-homogeneous case.

Definition A.18. *We say that a time-homogeneous F-doubly stochastic Markov chain X has an intensity $\boldsymbol{\Psi} = [\psi_{j,k}]_{j,k\in\mathcal{K}}$ if $\boldsymbol{\Psi}$ is $\mathcal{F}_T$-measurable and satisfies the following conditions*

(i) $0 \leq \psi_{j,k} < \infty$ $\mathbb{P}$-a.s. for all $j,k \in \mathcal{K}$ with $j \neq k$,

(ii) $\psi_{j,j} = -\sum_{\substack{k=1\\k\neq j}}^{N} \psi_{j,k}$ $\mathbb{P}$-a.s. for all $j \in \mathcal{K}$,

(iii) $\boldsymbol{\Psi}$ solves $\frac{d\boldsymbol{P}(t)}{dt} = \boldsymbol{\Psi}\boldsymbol{P}(t)$, $\boldsymbol{P}(0) = \boldsymbol{I}_N$ (Kolmogorov backward equation) and $\boldsymbol{\Psi}$ solves $\frac{d\boldsymbol{P}(t)}{dt} = \boldsymbol{P}(t)\boldsymbol{\Psi}$, $\boldsymbol{P}(0) = \boldsymbol{I}_N$ (Kolmogorov forward equation),

where $\boldsymbol{I}_N$ is the $N \times N$ identity matrix.

From Proposition A.19 we obtain the following proposition.

Proposition A.19. *Let X be a time-homogeneous F-doubly stochastic Markov chain with intensity $\boldsymbol{\Psi}$ and let $(\tau_k)_{k\geq0}$ be the sequence of jump-times given by (A.15). Provided that $\tau_k < \infty$ $\mathbb{P}$-a.s., we have for $t \in \mathbb{R}_+$*

$$\mathbb{P}(\tau_k - \tau_{k-1} > t\,|\,\widetilde{\mathcal{G}}_{\tau_{k-1}}) = e^{-t\psi_{X_{\tau_{k-1}}}}, \tag{A.24}$$

where for all $k \in \mathcal{K}$ we denote $\psi_k := \psi_{k,k}$.

B. Galtchouk-Kunita-Watanabe Decompositions

We comment now on different versions of GKW-decompositions, their connection, and when they can be used for the quadratic hedging frameworks presented in Subsection 1.1.2. To this end, we assume given an $\mathbb{R}^d$-valued local martingale on $(\Omega, \mathcal{G}, \mathbb{G}, \mathbb{P})$, which we denote by $\widehat{\mathbf{S}}$ in accordance to Section 1.1.2.

Recall the set

$$L^2(\widehat{\mathbf{S}}) := \left\{ \boldsymbol{\psi} \,\middle|\, \boldsymbol{\psi} \text{ predictable}, \mathbb{E}\left[\int_0^T \boldsymbol{\psi}_s^\mathsf{T} d[\widehat{\mathbf{S}}]_s \boldsymbol{\psi}_s \right]^{\frac{1}{2}} < \infty \right\}.$$

For every $\boldsymbol{\psi} \in L^2(\widehat{\mathbf{S}})$, the process $(\boldsymbol{\psi} \cdot \widehat{\mathbf{S}})$ is a square-integrable martingale and we denote by

$$\mathcal{I}^2(\widehat{\mathbf{S}}) := \left\{ \left(\int_0^t \boldsymbol{\psi}_s^\mathsf{T} d\widehat{\mathbf{S}}_s \right)_{t \in [0,T]} : \boldsymbol{\psi} \in L^2(\widehat{\mathbf{S}}) \right\} \subseteq \mathcal{M}_0^2$$

the set of all integral processes with respect to $\widehat{\mathbf{S}}$ with integrands in $L^2(\widehat{\mathbf{S}})$. It is well known that $\mathcal{I}^2(\widehat{\mathbf{S}})$ is a closed stable subset of $\mathcal{M}_0^2$, the set of square-integrable martingales null at 0.

Definition B.1. *1) Given a square integrable martingale $U = (U_t)_{t \in [0,T]} \in \mathcal{M}^2$, the* Galtchouk-Kunita-Watanabe decomposition *of U with respect to $\widehat{\mathbf{S}}$ is given as*

$$U_t = U_0 + \int_{0+}^t (\boldsymbol{\varphi}_s^U)^\mathsf{T} d\widehat{\mathbf{S}}_s + L_t^U, \quad t \in [0,T], \tag{B.1}$$

where $\boldsymbol{\varphi}^U \in L^2(\widehat{\mathbf{S}})$ and L^U a square-integrable martingale null at 0 which is strongly orthogonal to the space $\mathcal{I}^2(\widehat{\mathbf{S}})$ in the sense that the product $(\boldsymbol{\phi} \cdot \widehat{\mathbf{S}})L^U$ is a local martingale for each $\boldsymbol{\phi} \in L^2(\widehat{\mathbf{S}})$.

2) Given a local martingale $N = (N_t)_{t \in [0,T]}$, the generalized Galtchouk-Kunita-Watanabe decomposition *of N with respect to $\widehat{\mathbf{S}}$ is given as*

$$N_t = N_0 + \int_{0+}^t (\boldsymbol{\varrho}_s^N)^\mathsf{T} d\widehat{\mathbf{S}}_s + \widetilde{L}_t^N, \quad t \in [0,T], \tag{B.2}$$

where $\boldsymbol{\varrho}^N \in L(\widehat{\mathbf{S}})$, i.e. $\boldsymbol{\varrho}^N$ is a predictable process such that the integral with respect to $\widehat{\mathbf{S}}$ exists, and $\widetilde{L}^N$ a local martingale null at 0 which is strongly orthogonal to $\widehat{\mathbf{S}}$ in the sense that for all $i \in \{1, ..., d\}$ the product $\widehat{S}^i L^N$ is a local martingale.

Remark B.2. *1) The existence of the GKW-decomposition (B.1) is guaranteed for each square integrable martingale U because the space $\mathcal{I}^2(\widehat{\mathbf{S}})$ is a closed stable subspace of $\mathcal{M}_0^2$, see also Schweizer [85].*

On the contrary, the generalized GKW-decomposition (B.2) *must not always exist, see Ansel and Stricker [4] for counterexamples.*

2) Both decompositions (if existing) are unique up to indistinguishability, see Ansel and Stricker [4], Kunita and Watanabe [68] or Schweizer [84].

Recall that, in order to derive the unique mean-variance or risk-minimizing hedging strategies for a square-integrable discounted T-contingent claim $\widehat{C}$ or a square-integrable discounted cumulative payment process $\widehat{D} = (\widehat{D}_t)_{t \in [0,T]}$, we need to derive the GKW-decomposition in the form (B.1) for their discounted intrinsic value processes $\widehat{U}^C$ or $\widehat{U}^D$, respectively[1].

Given the square-integrable discounted cumulative payment process $\widehat{D}$ in a concrete setting as e.g. in Chapters 3 and 4, it is often possible to derive a generalized GKW-decomposition for $\widehat{U}^D$ according to (B.2) but it is a priori not apparent that the obtained decomposition constitutes also the GKW-decomposition according to (B.1).

Recall e.g. Chapter 3 where the square-integrable discounted value process $\widehat{U}^{pe}$ of a pure endowment is decomposed by

$$\widehat{U}^{pe}_t = c^{pe} + \int_0^t \alpha_s^W dW_s + \underbrace{\int_{0+}^t \alpha_s^M dM}_{=: \widetilde{L}^{pe}} . \tag{B.3}$$

The martingales M and W are strongly orthogonal such that $\widetilde{L}^{pe}$ is strongly orthogonal to W as well. As the integrands α^W and α^M are predictable, (B.3) constitutes the generalized GKW-decomposition of $\widehat{U}^{pe}$ with respect to W. Yet, we cannot infer without further argumentation that (B.3) also constitutes the GKW-decomposition (B.1) for $\widehat{U}^{pe}$ with respect to W, because we do a priori neither know if $\alpha^W \in L^2(W)$ nor if $\widetilde{L}^{pe} \in \mathcal{M}_0^2$, i.e. $\alpha^M \in L^2(M)$.

Note that the integrability conditions $\alpha^W \in L^2(W)$ or $\alpha^M \in L^2(M)$ may in some cases be difficult to prove. Fortunately, the following Lemma provides a condition under which the GKW-decomposition and the generalized GKW-decomposition coincide.

Lemma B.3. *Provided there exists a GKW-decomposition* (B.1) *and a generalized GKW-decomposition* (B.2) *for a square-integrable martingale U with respect to $\widehat{S}$, then the two decompositions coincide if $\widehat{S} \in \mathcal{M}_0^2$.*

Proof. As $\widehat{S}$ is a square-integrable martingale null at 0, we have $\widehat{S} \in \mathcal{I}^2(\widehat{S})$. Hence, decomposition (B.1) provides that the square-integrable martingale L^U is strongly orthogonal to $\widehat{S}$. The uniqueness of the generalized GKW-decomposition (B.2) then proves the assertion. $\square$

---

[1] As every T-contingent claim C can be defined as a cumulative payment process, we continue by only considering $\widehat{D}$.

Considering again the pure endowment example, we have that $W \in \mathcal{M}_0^2$. Hence, Lemma B.3 shows that the decomposition (B.3) is the GKW-decomposition of $\widehat{U}^{pe}$ with respect to W. In particular we have $\left(\int_{0+}^t \alpha_s^M dM_s\right)_{t \in [0,T]} = \widetilde{L}^{pe} \in \mathcal{M}_0^2$.

Recall that the discounted intrinsic value process $\widehat{U}^{lb}$ of the longevity bond which serves as hedging instrument in Chapter 3 has the decomposition

$$\widehat{U}_t^{lb} = c^{lb} + \int_0^t \xi_s dW_s \,.$$

This particularly implies $\mathcal{I}^2(\widehat{U}^{lb}) \subseteq \mathcal{I}^2(W)$.

Moreover, we know that the GKW-decomposition

$$\widehat{U}_t^{pe} = c^{pe} + \int_0^t \vartheta_s^{pe} d\widehat{U}_s^{lb} + \widehat{L}^{pe} \tag{B.4}$$

$$= c^{pe} + \int_0^t \vartheta_s^{pe} \xi_s dW_s + \widehat{L}^{pe} \tag{B.5}$$

with $\vartheta^{pe} \in L^2(\widehat{U}^{lb})$ and $\widehat{L}^{pe} \in \mathcal{M}_0^2$ strongly orthogonal to the space $\mathcal{I}^2(\widehat{U}^{lb}) \subseteq \mathcal{I}^2(W)$ exists. The uniqueness of this decomposition and (B.3) then show that $\left(\int_{0+}^t \alpha_s^M dM_s\right)_{t \in [0,T]} = L^{pe} = \widehat{L}^{pe}$ and that ϑ^{pe} is determined uniquely by

$$\vartheta^{pe} \xi = \alpha_s^W \,.$$

References

[1] O. O. Aalen, J. Fosen, H. Weedon-Fekjær, Ø. Borgan, and E. Husebye. Dynamic analysis of multivariate failure time data. *Biometrika*, 60:764–773, 2004.

[2] P. K. Andersen, Ø. Borgan, R. D. Gill, and N. Keiding. *Statistical Models Based on Counting Processes*. Springer Series in Statistics, Springer, New York, 2nd edition, 1993.

[3] J. P. Ansel and C. Stricker. Lois de martingale, densités et décomposition de Föllmer Schweizer. In *Annales de l'institut Henri Poincaré (B), Probabilités et Statistiques*, volume 28, pages 375–392. 1992.

[4] J. P. Ansel and C. Stricker. Décomposition de Kunita-Watanabe. In *Séminaire de Probabilités XXVII*, Lecture Notes in Mathematics Volume 1557. Springer, 1993.

[5] J. Barbarin. *Valuation, Hedging and the Risk Management of Insurance Contracts*. PhD thesis, Université Catholique de Louvain, 2008.

[6] J. Barbarin. Heath-Jarrow-Morton modelling of longevity bonds and the risk minimization of life insurance portfolios. *Insurance: Mathematics and Economics*, 43(1): 41–55, 2008.

[7] P. Barrieu and L. Albertini, editors. *The Handbook of Insurance-Linked Securities*. Wiley, Chichester, 2009.

[8] D. Becherer. The numeraire portfolio for unbounded semimartingales. *Finance and Stochastics*, 5:327–341, 2001.

[9] F. Biagini. Evaluating hybrid products: the interplay between financial and insurance markets. In R. Dalang, M. Dozzi, and F. Russo, editors, *Seminar on Stochastic Analysis, Random Fields and Applications VII, Centro Stefano Franscini, Ascona, May 2011*. Springer, 2014. To appear.

[10] F. Biagini and A. Cretarola. Quadratic hedging methods for defaultable claims. *Applied Mathematics and Optimization*, 56(3):425–443, 2007.

[11] F. Biagini and A. Cretarola. Local risk-minimization for defaultable markets. *Mathematical Finance*, 19:669–689, 2009.

[12] F. Biagini and A. Cretarola. Local risk-minimization for defaultable claims with recovery process. *Applied Mathematics & Optimization*, 65:293–314, 2012.

[13] F. Biagini and I. Schreiber. Risk minimization for life insurance liabilities. *SIAM Journal on Financial Mathematics*, 2013. In press.

[14] F. Biagini and J. Widenmann. Pricing of unemployment insurance products with doubly-stochastic Markov chains. *International Journal of Theoretical and Applied Finance*, 15(4):1250025 (32 pages), 2012.

[15] F. Biagini and J. Widenmann. Risk minimization for insurance products via F-doubly stochastic Markov chains. Working Paper, University of Munich, 2013.

[16] F. Biagini, A. Cretarola, and E. Platen. Local risk-minimization under the benchmark approach. Working Paper, University of Munich, Universitá degli Studi di Perugia, University of Technology Sydney, 2011.

[17] F. Biagini, C. Botero, and I. Schreiber. Risk-minimization for life insurance liabilities with dependent mortality risk. Working Paper, University of Munich, 2012.

[18] F. Biagini, T. Rheinländer, and I. Schreiber. Risk-minimization for life insurance liabilities with basis risk. Working Paper, University of Munich and TU Vienna, 2012.

[19] F. Biagini, A. Groll, and J. Widenmann. Intensity-based premium evaluation for unemployment insurance products. *Insurance: Mathematics and Economics*, 53(1): 302–316, 2013.

[20] F. Biagini, T. Rheinländer, and J. Widenmann. Hedging mortality claims with longevity bonds. *ASTIN Bulletin*, 43(2):123–157, 2013.

[21] T. R. Bielecki, M. Jeanblanc, and M. Rutkowski. Pricing and hedging of credit risk: Replication and mean-variance approaches. I. In G. Yin and Q. Zhang, editors, *Mathematics of Finance*, pages 37–53. Contemporary Mathematics, 351, AMS, Providence, 2004.

[22] T. R. Bielecki, M. Jeanblanc, and M. Rutkowski. Pricing and hedging of credit risk: Replication and mean-variance approaches II. In G. Yin and Q. Zhang, editors, *Mathematics of Finance*, pages 55–64. Contemporary Mathematics, 351, AMS, 2004.

[23] T. R. Bielecki, M. Jeanblanc, and M. Rutkowski. Pricing and trading credit default swaps in a hazard process model. *Annals of Applied Probability*, 18(6):2495–2529, 2008.

[24] T.R. Bielecki and M. Rutkowski. *Credit Risk: Modelling, Valuation and Hedging*. Springer-Finance. Springer, second edition, 2004.

[25] E. Biffis. Affine processes for dynamic mortality and actuarial valuation. *Insurance: Mathematics and Economics*, 37(3):443–468, 2005.

[26] E. Biffis and P. Millossovich. A bidimensional approach to mortality risk. *Decisions in Economics and Finance*, 29(2):71–94, 2006.

[27] E. Biffis, M. Denuit, and P. Devolder. Stochastic mortality under measure changes. *Scandinavian Actuarial Journal*, 2010:284–311, 2010.

[28] F. Black and M. Scholes. The pricing of options and corporate liabilities. *Journal of political economy*, 81(3):637–654, 1973.

[29] D. Blake and W. Burrows. Survivor bonds: helping to hedge mortality risk. *Journal of Risk and Insurance*, 68(2):336–348, 2001.

[30] D. Blake, A. J. G. Cairns, and K. Dowd. Pricing death: frameworks for the valuation and securitization of mortality risk. *ASTIN Bulletin*, 36(1):79–120, 2006.

[31] D. Blake, A. J. G. Cairns, and K. Dowd. The birth of the life market. *Asia-Pacific Journal of Risk and Insurance*, 3(1):6–36, 2008.

[32] D. Blake, T. Boardman, and A. J. G. Cairns. Longevity risk: Why governments should issue longevity bonds. Discussion Paper. Pensions Institute, 2010.

[33] C. Blanchet-Scalliet and M. Jeanblanc. Hazard rate for credit risk and hedging defaultable contingent claims. *Finance & Stochastics*, 8(1):145–159, 2004.

[34] N. Bouleau and D. Lamberton. Residual risks and hedging strategies in markovian markets. *Stochastic Processes and their Applications*, 33:131–150, 1989.

[35] P. Brémaud and M. Yor. Changes of filtration and of probability measures. *Z. Wahrscheinlichkeitstheorie verw. Gebiete*, 45:269–295, 1978.

[36] D. Coculescu, M. Jeanblanc, and A. Nikeghbali. Default times, non arbitrage conditions and change of probability measures. Working Paper, University of Zurich, Université d'Evry, 2009.

[37] D. R. Cox. Regression models and life tables. *Journal of the Royal Statistical Society. Series B (Methodological)*, 34(2):187–220, 1972.

[38] D. R. Cox. Partial likelihood. *Biometrika*, 62(2):269–276, 1975.

[39] D. R. Cox and E. J. Snell. A general definition of residuals (with discussion). *Journal of the Royal Statistical Society series B*, 30:248–275, 1968.

[40] M. Dahl and T. Møller. Valuation and hedging of life insurance liabilities with systematic mortality risk. *Insurance: Mathematics and Economics*, 39(2):193–217, 2006.

[41] M. Dahl, M. Melchior, and T. Møller. On systematic mortality risk and risk-minimization with survivor swaps. *Scandinavian Actuarial Journal*, 2(2):114–146, 2008.

[42] L. C. de Wreede, M. Fiocco, and H. Putter. The mstate package for estimation and prediction in non- and semi-parametric multi-state and competing risks models. *Computer Methods and Programs in Biomedicine*, 99:261–274, 2010.

[43] F. Delbaen and J. Haezendonck. A martingale approach to premium calculation principles in an arbitrage free market. *Insurance: Mathematics and Economics*, 8(4): 269–277, 1989.

[44] F. Delbaen and W. Schachermayer. A general version of the fundamental theorem of asset pricing. *Mathematische Annalen*, 300:463–520, 1994.

[45] F. Delbaen and W. Schachermayer. The fundamental theorem of asset pricing for unbounded stochastic processes. *Mathematische Annalen*, 312:215–250, 1998.

[46] H.M. Dietz. On the solution of matrix-valued linear stochastic differential equations driven by semimartingales. *Stochastics and Stochastic Reports*, 34(3–4):127 – 147, 1991.

[47] M. Dorner, M. König, and S. Seth. FDZ-Datenreport: Sample of integrated labour market biographies; regional file 1975-2008 (siab-r 7508). Technical report, Research Data Centre fo the German Federal Employment Agency (BA) and the Institut for Employment Research (IAB), 2011.

[48] D. Duffie. *Dynamic Asset Pricing Theory*. Princeton University Press, 3 edition, 2001.

[49] D. Duffie, J. Pan, and K. Singleton. Transform analysis and asset pricing for affine jump-diffusions. *Econometrica*, 68(6):1343–1376, 2000.

[50] D. Duffie, D. Filipović, and W. Schachermayer. Affine processes and applications in finance. *Annals of Applied Probability*, 13(3):984–1053, 2003.

[51] E.R. Fernholz. *Stochastic Portfolio Theory*. Springer New York, 2002.

[52] E.R. Fernholz and I. Karatzas. Stochastic portfolio theory: an overview. In A. Bensoussan and Q. Zhan, editors, *Handbook of Numerical Analysis; Volume XV "Mathematical Modeling and Numerical Methods in Finance"*, pages 89–167. North Holland, 2009.

[53] H. Föllmer and A. Schied. *Stochastic finance*. De Gruyter, 2004.

[54] H. Föllmer and D. Sondermann. Hedging of non-redundant contingent claims. In A. Mas-Colell and W. Hildenbrand, editors, *Contributions to Mathematical Economics*, pages 205–223. North Holland, 1986.

[55] C. Fontana and W. J. Runggaldier. Diffusion-based models for financial markets without martingale measures. Working Paper, University of Padova, 2011.

[56] P. Grambsch and T. Therneau. Proportional hazards tests and diagnostics based on weighted residuals. *Biometrika*, 81:515–526, 1994.

[57] M. Harrison and S. Pliska. Martingales and stochastic integrals in the theory of continuous trading. *Stochastic Processes and their Applications*, 11(3):215–260, 1981.

[58] H. Hulley and M. Schweizer. M^6 - on minimal market models and minimal martingale measures. In C. Chiarella and A. Novikov, editors, *Contemporary Quantitative Finance. Essays in Honour of Eckhard Platen*, pages 35–51. Springer, 2010.

[59] K. Ignatieva and E. Platen. Estimating the diffusion coefficient function for a diversified world stock index. *Computational Statistics and Data Analysis*, 56:1333–1349, 2012.

[60] C. H. Jackson. Multi-state models for panel data: The msm package for r. *Journal of Statistical Software*, 38(8):1–29, 2011.

[61] J. Jakubowski and M. Niewęgłowski. A class of F-doubly stochastic markov chains. *Electronic journal of probability*, 15:1743–1771, 2010.

[62] J. Jakubowski and M. Niewęgłowski. Pricing and hedging of rating-sensitive claims modeled by F-doubly stochastic Markov chains. In G. diNunno and Bernt Øksendal, editors, *Advanced Mathematical Methods for Finance*, pages 417 – 454. Springer, 2011.

[63] Y. Kabanov and D. Kramkov. No-arbitrage and equivalent martingale measures: an elementary proof of the harrison-pliska theorem. *Theory of Probability and its Applications*, 39(3):523–527, 1994.

[64] J. D. Kalbfleisch and J. F. Lawless. The analysis of panel data under a markov assumption. *Journal of the American Statistical Association*, 80(392):863–871, 1985.

[65] I. Karatzas and C. Kardaras. The numéraire portfolio in semimartingale financial models. *Finance and Stochastics*, 11(4):447–493, 2007.

[66] T. Kneib and A. Hennerfeind. Bayesian semiparametric multi-state models. *Statistical Modelling*, 8(2):169–198, 2008.

[67] A. Kull. A unifying approach to pricing insurance and financial risk. Working Paper, Casualty Actuarial Society, 2003.

[68] H. Kunita and S. Watanabe. On square integrable martingales. *Nagoya Mathematical Journal*, 30:209–245, 1967.

[69] E. Luciano and E. Vigna. Mortality risk via affine stochastic intensities: calibration and empirical relevance. *Belgian Actuarial Bulletin*, 8:5–16, 2008.

[70] R.C. Merton. Theory of rational option pricing. *The Bell Journal of Economics and Management Science*, 4(1):141–183, 1973.

[71] T. Møller. Risk-minimizing hedging strategies for unit-linked life insurance contracts. *ASTIN Bulletin*, 28(1):17–47, 1998.

[72] T. Møller. Risk-minimizing hedging strategies for insurance payment processes. *Finance and Stochastics*, 5(4):419–446, 2001.

[73] R. Norberg. Hattendorff's theorem and Thiele's differential equation generalized. *Scandinavian Actuarial Journal*, 1:2–14, 1992.

[74] R. Norberg. The pension crisis: its causes, possible remedies, and the role of the regulator. *Erfaringer og utfordringer, 20 years Jubilee Volume of Kredittilsynet, the Financial Supervisory Authority of Norway*, 2006.

[75] H. Pham. On quadratic hedging in continuous time. *Mathematical Methods of Operations Research*, 51:315–339, 2000.

[76] E. Platen. A benchmark framework for risk management. In *Stochastic Processes and Applications to Mathematical Finance, Proceedings of the Ritsumeikan Intern. Symposium*, pages 305–335, 2004.

[77] E. Platen. Diversified portfolios with jumps in a benchmark framework. *Asia-Pacific Financial Markets*, 11(1):1–22, 2005.

[78] E. Platen and D. Heath. *A Benchmark Approach to Quantitative Finance*. Springer, 2007.

[79] E. Platen and R. Rendek. Simulation of diversified portfolios in a continuous financial market. QFRC Working Paper, University of Sydney, 2011.

[80] P. E. Protter. *Stochastic Integration and Differential Equations*. Springer, New York, second edition, 2005.

[81] T. Rolski, H. Schmidli, V. Schmidt, and J. Teugels. *Stochastic Processes for Insurance and Finance*. John Wiley & Sons, 1999.

[82] D. Schoenfeld. Partial residuals for the proportional hazards regression model. *Biometrika*, 69:239–241, 1982.

[83] D. F. Schrager. Affine stochastic mortality. *Insurance: Mathematics and Economics*, 38(1):81–97, 2006.

[84] M. Schweizer. On the minimal martingale measure and the Föllmer-Schweizer decomposition. *Stochastic Analysis and Applications*, 13:573–599, 1995.

[85] M. Schweizer. A guided tour through quadratic hedging approaches. In E. Jouini, J. Cvitanic, and M. Musiela, editors, *Option Pricing, Interest Rates and Risk Management*, pages 538–574. Cambridge University Press, Cambridge, UK, 2001.

[86] M. Schweizer. From actuarial to financial valuation principles. *Insurance: Mathematics and Economics*, 28:31–47, 2001.

[87] M. Schweizer. Local risk-minimization for multidimensional assets and payment streams. *Banach Center Publications*, 83:213–229, 2008.

[88] N. Simon, J. Friedman, T. Hastie, and R. Tibshirani. Regularization paths for Cox's proportional hazards model via coordinate descent. *Journal of Statistical Software*, 39 (5):1–13, 2011.

[89] D. Sondermann. Reinsurance in arbitrage-free markets. *Insurance: Mathematics and Economics*, 10:191–202, 1991.

[90] K. Takaoka. A note on the condition of no unbounded profit with bounded risk. *Finance and Stochastics*, 2012. In press.

[91] T. Therneau. *A Package for Survival Analysis in S*, 2012. R package version 2.36-14.

[92] T. Therneau and P. Grambsch. *Modeling Survival Data: Extending the Cox Model*. Springer-Verlag, 2000.

[93] N. Vandaele and M. Vanmaele. A locally risk-minimizing hedging strategy for unit-linked life insurance contracts in a lévy process financial market. *Insurance: Mathematics and Economics*, 42(3):1128–1137, 2008.

[94] M. Wadsworth, A. Findlater, and T. Boardman. Reinventing annuities. Working Paper, Staple Inn Society, 2001.

[95] C. Weber. *Insurance Linked Securities*. Gabler, 2011.